A FEW OPINIONS.

"The processes of assaying are detailed with great clearness. —*Popular Science Monthly, July,* 1883.

"It is just such a book as is needed to assist those who are, by actual work developing the gold and silver mines of the country."—*Chicago Tribune, March* 31, 1883.

"The author has faithfully endeavored to carry out his idea of a plain, practical work. No one will be a loser who possesses it."—*Mining and Scientific Press, San Francisco, Aug.* 18, 1883.

Has endeavored, we may add, with marked success, to prepare a work on assaying."—*Engineering and Mining Journal, April* 28, 1883.

"The book will form an excellent elementary guide to the important art of assaying."—*Chemical News, London, June,* 8, 1883.

"Mr. Brown has succeeded in providing a valuable assistant for the student, the miner and the assayer.—*Mining Review, March* 29, 1883.

"This handy volume appears to contain sufficient information in the art of assaying to entitle it to the claim of being a manual of instruction to beginners."—*Scientific American, July* 14, 1883.

"There is probably no other book in the English language that so fully meets the wants of those for whom it was made as this manual."—*Journal of Education, Boston, Sept.* 25, 1884.

"It is a complete compendium of everything pertaining to the assaying of the above mentioned ores."—*Mining Record, April* 7, 1883.

FORT MCGINNIS, MONTANA. July 22, 1886.
"Allow me, as a miner, to thank you for your manual. It is written in English, and for 'us,' and not for Freiberg men."—EUGENE SMITH.

PORTLAND, ORE., March 15, 1886.
"My opinion is, that for clearness, accuracy, and adaptation to the uses of an American assayer, it is worth all the other assay books that I am acquainted with."—H. O. LANG.

"LAWRENCE, KAS , July 22, 1884
"Your book is plain and to the point, and just the thing for beginners."—E. W. WALTER.

SAN FRANCISCO, CAL., Aug. 30, 1886.
"I congratulate you upon this little book. I wish other writers on metallurgical subjects would write in as plain English."—LOUIS JANIN, *Mining Engineer.*

"CHICAGO, April 3, 1883.
"The most essential point that struck me was the plain, common-sense manner in which you handle the subject. If more of our scientists would follow where you have dared to lead, and would free their subjects from useless technicalities and abstruse phrases, there would be less stumbling and misunderstanding among their students and readers, and more confidence given to them by those who have need of their services."—H. D. WHITTEMORE, C. E. and M. E.

MANUAL . OF . ASSAYING

GOLD . SILVER . COPPER . AND . LEAD

ORES . BY . WALTER . LEE . BROWN . B. Sc.

ONE . COLORED . PLATE . AND . ONE . HUNDRED
AND . THIRTY . TWO . ILLUSTRATIONS . ON . WOOD

FIFTH . EDITION . AND . FIFTH . THOUSAND . . . .

"ALL . IS . NOT . GOLD . THAT . OUTWARD . SHEWTH . BRIGHT"

CHICAGO . PUBLISHED . BY . E . H . SARGENT . &
CO . EIGHTEEN . HUNDRED - AND . NINETY . FOUR

EXTRACT FROM PREFACE TO FIRST EDITION.

WHEN I entered upon the task of preparing this book, it was with the idea of furnishing a guide to those, who, having had no previous technical or especially scientific education, desired to learn something of . the *practical assaying of gold and silver ores*, and in whose hands I could place no work that could give them this information in a clear, simple and thoroughly detailed manner, and unburdened with unnecessary matter.

This intention I have tried to adhere to all the way through, and while I have added other information which was pertinent, I have kept such increase in an appendix, so that the body proper of the work contains the real subject matter.

It is my sincere belief that there is no book in the English language on the subject of assaying which occupies the space that this little manual tries to fill.

A number of such publications fail to meet the want, on account of their antiquity, they having been written some thirty years ago ; hence their methods, apparatus, etc., are not suited to the assayers of to-day. Others are either more suitable as books of reference, or do not give sufficient detail for the inexperienced.

It is this latter fault I have carefully endeavored to avoid, and perhaps have gone to the other extreme. At all events, I have tried to give here in print the precise instruction which I have previously imparted orally to my students. Those who may choose to criticise, will remember for whom this hand-book is written.

I wish to publicly thank the following gentlemen who have very kindly aided me in my work with valuable information: Mr. S. A. Reed, Irwin, Colo.; Mr. A. H. Low, Assayer of the Boston and Colorado Smelting Works, Argo, Colo.; Mr. C. Boyer, Assayer of the U. S. Branch Mint, Denver Colo.; Mr. M. G. Nixon, Engineer, and Mr. R. G. Coates, both of Chicago.

March, 1883.

PREFACE TO THE SECOND EDITION.

So many words of praise, both from the press and private individuals, greeted the first edition of this work, that I feel encouraged to put forth a second and greatly improved edition, which will not contain the faults that have been pointed out to me as existing in the first.

The salient features of the new comer are increase in matter from 318 to 488 pages; the stating of all charges in assay tons, grammes, and grains; detailed charges in the scorification process; full notes on the colors and appearances of the scorifiers (with a colored plate) and cupels after work; the expansion of the crucible process from *nine* to almost *ninety* pages; more complete articles on the assay of gold and silver bullion, and the volumetric analysis of copper ores; and, finally, the issuance of the book in flexible covers.

For the idea of the oxidizing powers of ores which I have developed in the notes on the crucible process, I am indebted to Aaron's Assaying.

Whatever appears in this edition extracted from the following authorities, is given with the *full and written permission of the authors, editors, or owners of the copyright*, as the case may be: Mitchell's Manual of Assaying, Kustel's Roasting of Gold and Silver Ores, Aaron's Assaying, Chapman's Furnace Assay, and Hank's Fourth Annual Report as the State Mineralogist of California.

Besides for that which I have incorporated from the above-mentioned sources, I wish to give thanks to the following gentlemen: to Mr. J. C. Jackson, my successor in business in Chicago, for illustrations of a permanent furnace, and particularly for the neat and characteristic frontispiece; to Mr. F. E. Fielding, Assayer, Virginia City, Nev., for valuable notes on the assay of gold bullion and volumetric analysis of copper ores; and to Mr. G. H. Ellis, for the careful working out of the qualitative schemes.

October 1, 1886.

NOTES TO THE THIRD EDITION.

This edition differs from the previous one only in the correction of a few typographical errors; in the revision (and reduction) of the prices of a number of articles listed in the assayer's outfit on pages 439-441, and some minor alterations in the descriptions of apparatus and re-agents.

It has been a matter of gratification to both publisher and author that at least one-fourth of the second edition has been sold in England, Germany and Australasia, and this without making any special effort to effect its sale in foreign countries.

July, 1889.

NOTES TO THE FOURTH EDITION.

In the course of time, all books, and especially technical works, become more or less obsolete, unless they keep pace with the improvements and progress of the age. So with this little work, it must be fully up to the times. Consequently the section on apparatus has been thoroughly revised, all matter, both illustrative and descriptive, representing apparatus not now commonly used, has been expunged, and new forms have been added. We have also reason to believe that the illustrations to this work have been considered as among its most important features, so that their number has been increased from ninety-four to one hundred and thirty-two.

Amongst the methods, some of those on "Lead Ores" have been re-written in accordance with the results of recent investigations. In the appendix, chapters have been added on the "Tin Assay," on "Gold and Silver Ores and Minerals," and a short chapter on the "Determination of the Specific Gravity of Minerals."

To fit the manual for the English market, where many copies of the previous editions have found a welcome, the pecuniary values of apparatus, etc., which have heretofore been given only in American dollars and cents, will now be found additionally in their English equivalents.

There has in general been a sharp revision, making good many little omissions, adding clearer explanations, and so on, all tending to bring this book a little nearer to that plane of perfection which we all seek.

January, 1892.

NOTES TO THE FIFTH EDITION.

Some changes will be observed in the chapter on apparatus, notably the insertion of the new "Ball Balance," "Kellogg Bunsen Vapor Lamp," the improved "Bosworth Crusher," etc.

The subject matter of the entire book has been critically examined for errors, mis-statements or other flaws, so that this edition may be as acceptable to the public as have been its predecessors.

March, 1894.   WALTER LEE BROWN.

CONTENTS.

PART I.

APPARATUS AND RE-AGENTS.

CHAPTER I.

APPARATUS USED IN ASSAYING.

CHAPTER II.

RE-AGENTS USED IN ASSAYING.

CHAPTER III.

TESTING OF RE-AGENTS.

PART II.

ASSAYING.

CHAPTER I.

. GOLD AND SILVER ORES.

CHAPTER II.

COPPER ORES.

CHAPTER III.

LEAD ORES.

APPENDIX.

SECTION I.

SPECIAL METHODS.

SECTION II.

LISTS AND REFERENCES.

SECTION III.

TABLES.

INTRODUCTION.

In our present state of knowledge we believe all matter to be composed of one or more elements or original simple substances.

These elements are considered to be seventy in number. Certain of them have what we may call a *commercial importance*. (See Introduction to Attwood's Blow-pipe Assaying.) They are as follows, the metals being in *italics:*

1. *Aluminium.*	16. *Gold.*
2. *Antimony.*	17. Hydrogen.
3. Arsenic.	18. Iodine.
4. *Barium.*	19. *Iridium.*
5. *Bismuth.*	20. *Iron.*
6. Boron.	21. *Lead.*
7 Bromine.	22. *Lithium.*
8. *Cadmium.*	23. *Magnesium.*
9. *Calcium.*	24. *Manganese.*
10. Carbon.	25. *Mercury.*
11. Chlorine.	26. *Molybdenum.*
12. *Chromium.*	27. *Nickel.*
13. *Cobalt.*	28. Nitrogen.
14. *Copper.*	29. Oxygen.
15. Fluorine.	30. *Palladium.*

31. Phosphorus.
32. *Platinum.*
33. *Potassium.*
34. Silicon.
35. *Silver.*
36. *Sodium.*
37. *Strontium.*
38. Sulphur.
39. *Tin.*
40. *Titanium.*
41. *Tungsten.*
42. *Uranium.*
43. *Vanadium.*
44. *Zinc.*
45. Zirconium.

Some of the above are valuable in themselves, others in combination.

The remainder of the elements, which have no especial value excepting perhaps as curiosities, are :

1. *Cæsium.*
2. *Cerium.*
3. *Columbium.*
4. *Davyum.*
5. *Decipium.*
6. *Didymium.*
7. *Erbium.*
8. *Gallium.*
9. *Glucinum.*
10. *Indium.*
11. *Lanthanum.*
12. *Norwegium.*
13. *Osmium.*
14. *Rhodium.*
15. *Rubidium.*
16. *Ruthenium.*
17. *Samarium.*
18. *Scandium.*
19. Selenium.
20. *Tantalum.*
21. Tellurium.
22. *Thallium.*
23. *Thorium.*
24. *Ytterbium.*
25. *Yttrium.*

Besides the seventy elements above enumerated, there are some ten or more extremely rare metals (actinium, gadolinium, germanium, helium, holmium, idunium, ilmenium, mosandrium, neptunium, philippium, terbium, thulium, etc.), whose existence is not yet quite satisfactorily proven.

In order to ascertain the value of an ore, it is necessary to determine the percentage of the metal or metals which it contains.

This is the first thing to be done—an after consideration is the question of the presence of other ingredients which may injuriously affect the value of the ore or product.

There are two general methods, known respectively as *assaying* and *analysis*, whereby we may test an ore to learn its composition.

A comprehensive definition of assaying is to call it that branch of exact science which enables us to find out of what a substance is composed and the proportions, by means of *dry* re-agents and *heat*.

On the other hand, analysis is that branch which effects the same results mainly by the use of *wet* re-agents, with or without the aid of heat.

In spite of this distinction, *wet assays*, as opposed to *dry* or *fire assays*, are continually spoken of ; still, to be as consistent as possible, the terms assaying and analysis, as defined above, will be used throughout this work.

The greater number of the processes given in this little book come under the former heading, while analysis proper is employed in only a few cases.

The following metals are sought for in ores by assaying : antimony, bismuth, cobalt, copper, gold, iron, lead, nickel, platinum, silver, tin and zinc.

It is the object of this manual to treat only of gold, silver, copper and lead. For information concerning the assaying of the remaining metals just mentioned, the student must seek it on page 434, among the various works on assaying there quoted.

PART I.

APPARATUS AND RE-AGENTS.

MANUAL OF ASSAYING.

PART I.

APPARATUS AND RE-AGENTS.

CHAPTER I.

APPARATUS USED IN ASSAYING.

It is as true of the art of assaying as of any other, that "good work requires good tools." While many of the latter can be dispensed with by the skilled assayer, it is often convenient, if not absolutely necessary, for the unskilled to have the best utensils for the work required.

I shall therefore give an exhaustive list of apparatus needed for the processes herein described, but shall try to avoid mentioning many implements which are not essential

IMPLEMENTS FOR PULVERIZING, SAMPLING, ETC.

Iron Mortars and Pestles.—Two sizes of mortars are handy, a large one 11 inches in diameter, and weighing with pestle about 35 pounds (2 gallons capacity), and a smaller one of 5 inches diameter and 7 pounds weight ($\frac{1}{9}$ gallon capacity). Instead of both, a medium size, 8 inches diameter, in weight about 19 pounds (1 gallon capacity), may be employed. They may be either bell or urn-shaped.

Fig. 1.

Care should be taken to remove all ore from the mortars after grinding. Generally an old towel, rag, or even paper, will suffice to do this, but occasionally washing must be resorted to. Dry thoroughly after the latter operation. Triturating with dry sand often answers the purpose. When not in use let the mortars rest mouth downward.

Fig. 2.

Mortars of various sizes can also be obtained with two oppositely situated projections, to serve either as handles for carrying, or as trunnions for turning over and dumping contents. Fig. 2 represents this pattern.

To ease the labor of lifting a heavy pestle, an arrangement similar to that shown in fig. 3 is recommended. This particular form is that which was in use in my own laboratory, where it always gave satisfaction.

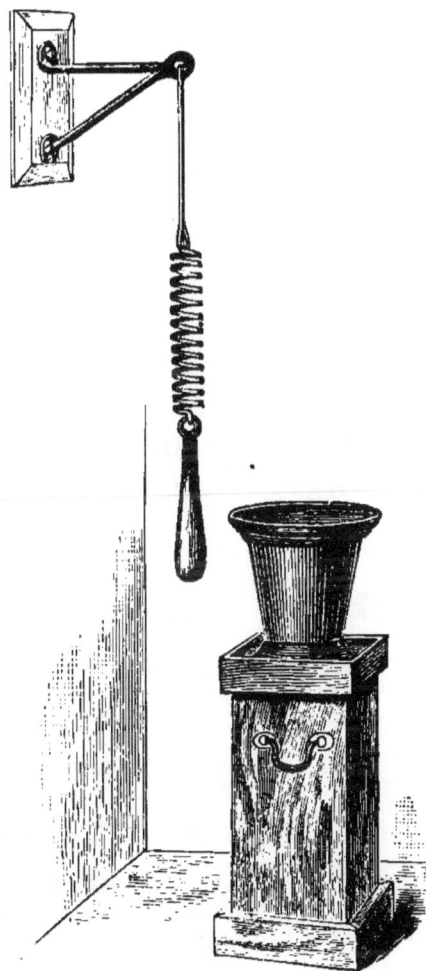

FIG. 3.

The spring does not perceptibly add to the force required to strike a crushing blow, but does materially aid in lifting the pestle. It (the spring) is 18 inches long (when unstretched) in coils of 1⅝ inches diameter, made of the best steel of ⅛ inch diameter, and painted with asphalt black as a protection against rust. It is connected to the bracket by a strong and flexible cord. The supporting bracket or hook is of ½ inch malleable iron. The distance of the eye from the wall I have made 13 inches, but it can, of course, vary to suit the circumstances. When not in use do not keep the spring taut, but let the lower end be hung loosely from a hook in the wall.

Instead of the above contrivance, a spring-board of hickory, 3 inches wide, ½ to 1 inch thick, and 10 feet long, the further end firmly fastened, the free end connecting by a strong rope with the pestle, will do.

Any assay laboratory, whether permanent or transient, will require at least one of the three sizes of mortars specified. For fixed laboratories, and where the quantity of ore to be crushed is considerable, a very large mortar, 18 inches deep and 12 inches wide (weight about 150 pounds), to stand on the floor, is very valuable. The pestle accompanying it is 3 feet long, and weighs about 18 pounds. The striking end is usually flattened out to a width of from 4 to 6 inches to cover more ore at a blow, and also to prevent the flying out of the crushed material. To use, the assayer stands over the mortar, grasping the pestle with both hands.

The spring or spring-board arrangement can likewise be applied to the pestle of this mortar.

To prevent pieces of ore which are being crushed in the mortar from flying out, it is a good plan to cut a piece of wood, pasteboard or tin of a circumference somewhat larger than that of the top of the mortar, with a hole in the centre large enough to admit of a little play to the pestle, and to lay this on top of the mortar while at work.

Crushers.—These are intended to take the place of the mortar and pestle for crushing comparatively large quantities of ores. For small samples the mortar will do very well,

but for say 20 pounds and upward, some sort
of a crusher will be a desideratum.   Especially will it be needed in a large assay laboratory.

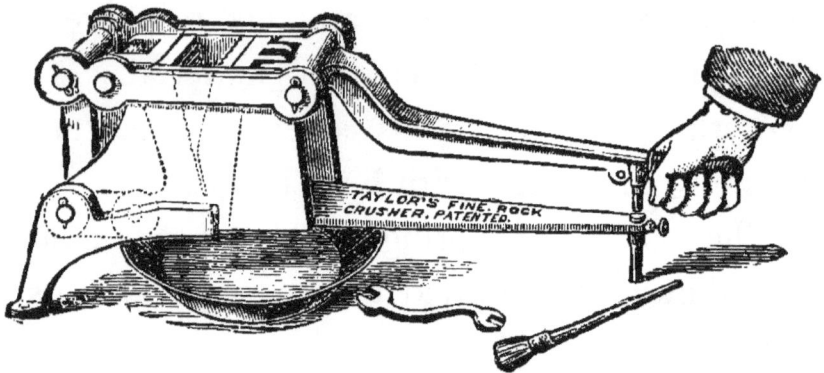

FIG. 4.

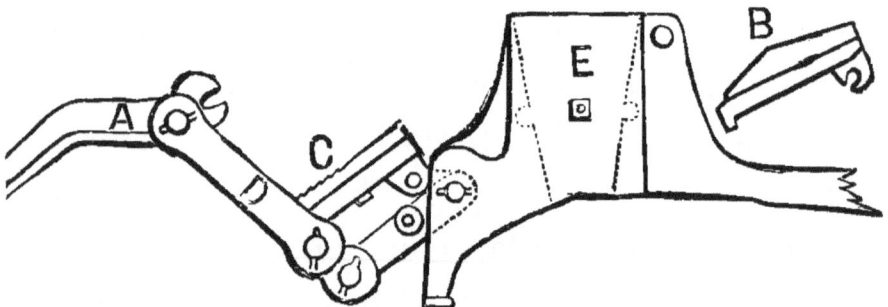

FIG. 5.

We are pleased to illustrate the "Taylor
Patent Rock Fine Crusher," in full in fig 4,
while fig 5 outlines the operating parts.

Both jaws are faced with hard white iron,

the lower parts of which are plain surfaces, between which the ore is crushed fine. The stationary jaw has its lower plain surface at an angle to the upper or corrugated surface. Lower part of this jaw is adjusted by a screw shown under the hand, to crush fine or coarse. The movable jaw is operated by the hand lever, and has its corrugations horizontal, to facilitate the forcing the ore down at each stroke of the lever. This jaw has a vertical and horizontal motion.

The lever has a rubber covering where grasped by the hand, and a rubber cushion where it strikes the bed-piece, to prevent jar and noise.

These jaws are 3 inches wide and open at the top 1¾ inches, consequently a piece of rock 3x1¾ inches can be crushed. With the lower part of jaws set at one-tenth inch apart, 40 pounds of the hardest rock can easily be crushed in one hour, and 20 per cent. of this will then go through a No. 60 sieve. Then screw up machine and soon run through the balance to No. 60 or finer. This machine crushes so much faster than the hand mortar and pestle, because of the great leverage and power, and because the fine crushed rock always drops away; whereas with mortar and pestle the fine is always in the way, and deadens each blow of the pestle.

This crusher can be used in crushing old crucibles, as well as for working up specimen ores, and is as equally useful to the prospector and sampler, as to the assayer. Its cost is $25.00 (£ 5, 3 sh.), and weight is 95 lbs.

For laboratories where a large quantity of ore is frequently to be crushed, I can heartily recommend the Gates crusher, shown in fig. 6. It requires power.

The operative mechanism of this machine, as seen in the engraving shown herewith, consists of a breaking head in the form of a frustrum or cone, placed vertically with the larger end of the cone toward the bottom or discharge point. The breaking head is

mounted upon a strong forged steel shaft, to
which motion is imparted through bevel gears
and an eccentric box. Power is transmitted
to the machine through the driving pulley seen
at the left of the picture. The surface of the
conical breaking head is either fluted, as seen

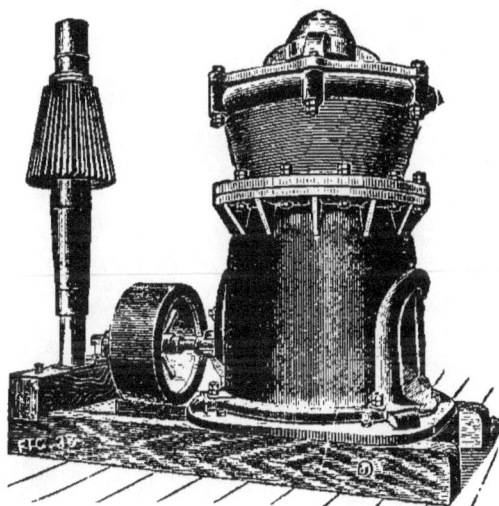

FIG. 6.

in the specimen head removed from the interior of the casing, or, when a very fine product is desired, the head is made smooth.
The interior face of the removable casing is
smooth, and the coarse material fed in from
above is continuously broken by being crushed

between the head and concave facing, and
the space between the crushing surfaces growing continuously smaller, it is discharged from
the bottom in fragments of any desired degree of fineness, the degree depending upon

FIG. 7.

the adjustment of the head, which adjustment is effected by a powerful screw at the
bottom of the shaft.

Fig 7 illustrates the Bosworth Crusher for
laboratory use, and which runs by either hand
or power.   It has been highly praised.

Pulverizing Plate and Rubbers.—These are so useful and convenient that they can hardly

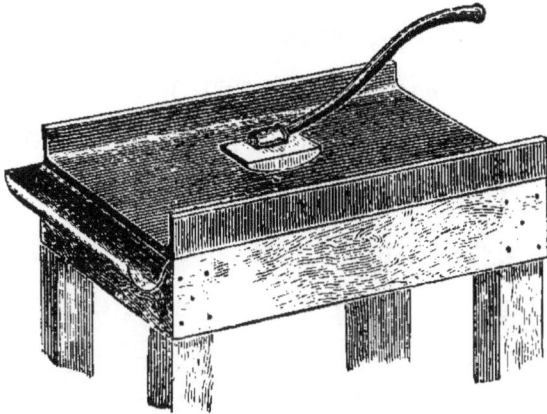

FIG. 8.

be dispensed with ; in fact, if much assaying is to be done, they will become absolutely necessary. They are represented in fig. 8. The iron plate, which should be perfectly true and have a smooth surface, is made of varying dimensions, as 12×12, 18×24, 23×24, 24×24, 24×30 and 24×36, in inches. It is made with no protecting rims, or with rims at the sides only (as figured), or with a rim at each side and at one end, leaving one end open. In some cases the side rims gradually shallow from the back to the front. The size I use, and with which I have no fault to find, is 18×24 inches (inside measurement); thickness of bottom, 1 inch; rims, $\frac{3}{4}$ inch wide and $1\frac{1}{4}$ inches high; weight, 150 pounds.

(The only improvement I can suggest would be to have the back corners rounded instead of being right angles, in order to facilitate the removal of pieces of ore and the dust. Or the rims might be cast at an angle of say 30° instead of being perpendicular to the surface.)

The rubber, rocker, pulverizer, grinder, muller, bucking-hammer (by which various names it is known), to go with above plate, is 8 inches long and 4 wide; thickness at ends, $1\frac{1}{8}$ inches; in centre, $2\frac{1}{8}$; surface true and smooth; weight about 14 pounds. Other rockers made are, in general dimensions, 4×5, 4×5½, 4×6, 6×7 and 8×10 inches.

An axe-handle is fitted into the socket on top of rocker, and then it is ready for use.

The operation of grinding, or rubbing, or

pulverizing is described under the treatment of the ores in Part II.

When a very hard ore is to be pulverized, it can much more quickly be finished with an 8×10 rocker, weight 60 pounds. The additional weight and greater width have a marked effect. An intermediate size of about 30 pounds' weight would not be amiss.

The plate can be placed upon a stout table, but it will be better in the long run to have constructed for it a special and substantial frame-work, as the long continued rubbing on the plate will eventually dislocate any ordinary table. The frame can have vertical legs as shown in the cut, or can have them set with an angle of a few degrees' spread at the floor to which they should be firmly screwed, using L-shaped plates of iron as the holders. Additional strength may be secured by inserting a thick shelf at about six inches from the floor, and this can serve as a resting place for the mortars where they will be out of the way.

In Mr. S. A. Reed's laboratory the plate is

set at an angle inclining toward the operator so as to allow of more effective pressure at the bottom, and slipped under the front end of the plate is a trough or gutter of sheet-tin or zinc, as shown in the figure. Its object is to catch any particles that may roll down the plate, and after the sample has been pulverized, the whole of the powder is brushed down into it. The trough is easily detached from the plate and its contents can then be brushed into a sieve. This simple device may replace one of the zinc sifting pans spoken of elsewhere.

Sample Shovels.—A pitch-fork with each tine transformed into a narrow trough would give a fair idea of the appearance of one of these shovels. A better form consists of two troughs from four to six inches deep united, with a space between, and provided with a long handle. These implements are more needed in sampling works proper than in the ordinary laboratory.

Samplers (also known as dividers). — One

form of these is shown in fig. 9,
and consists of a frame with
partitions running lengthwise
at equal distances apart, and

Fig. 9.

having each alternate space covered at the
bottom. It is made of tin or copper, and it
is well to have three sizes. A second pattern
is represented in
fig. 10. The first
form, however, is
preferred, as being
more durable.

Fig. 10.

To use either, sprinkle over and across the
broken ore to be sampled, and retain that
which catches in the troughs,

A pulp is sometimes sampled by the use of
a sampler of tin, having troughs $\frac{1}{8}$ inch in diameter and $\frac{3}{16}$ inch apart. To use, sprinkle the
sample across, over a piece of clean paper,
and separate that which goes between the
troughs from that which catches in them. After putting aside the latter portion, sprinkle
across the sampler the former portion, and so

continue until a quantity is obtained about
sufficient for assay.

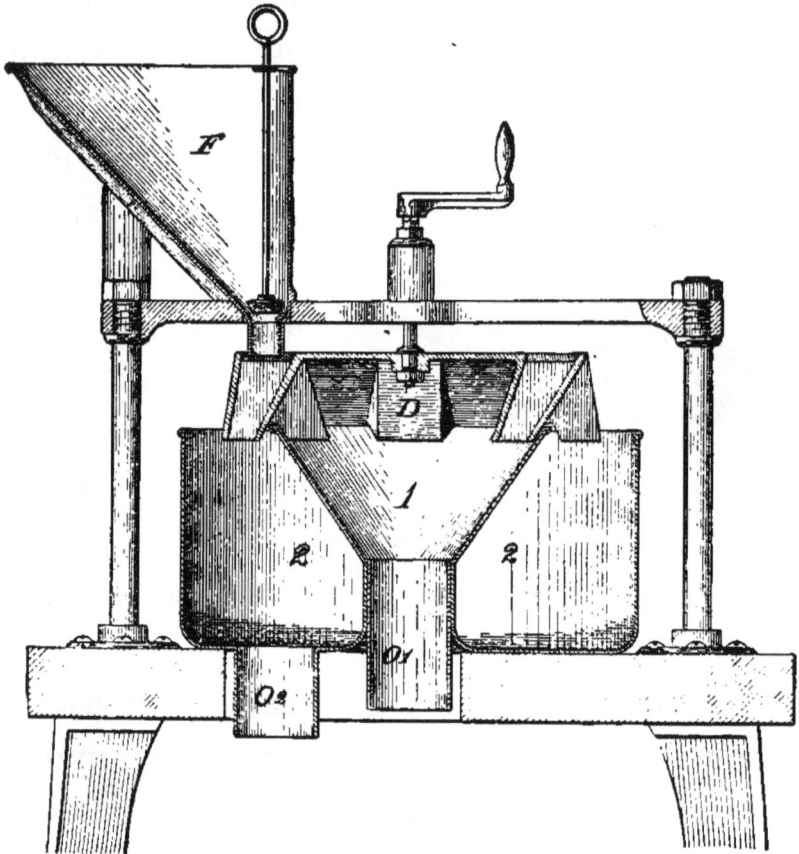

Fig. 11.

Ore Sampling Machines.—We are pleased
to illustrate and describe the ore sampling
machine invented by Mr. H. L. Bridgman,

for use in assay offices, sampling and smelting works, stamp mills, etc.

This machine is a modification of the large one, which is doing excellent work. Its par-

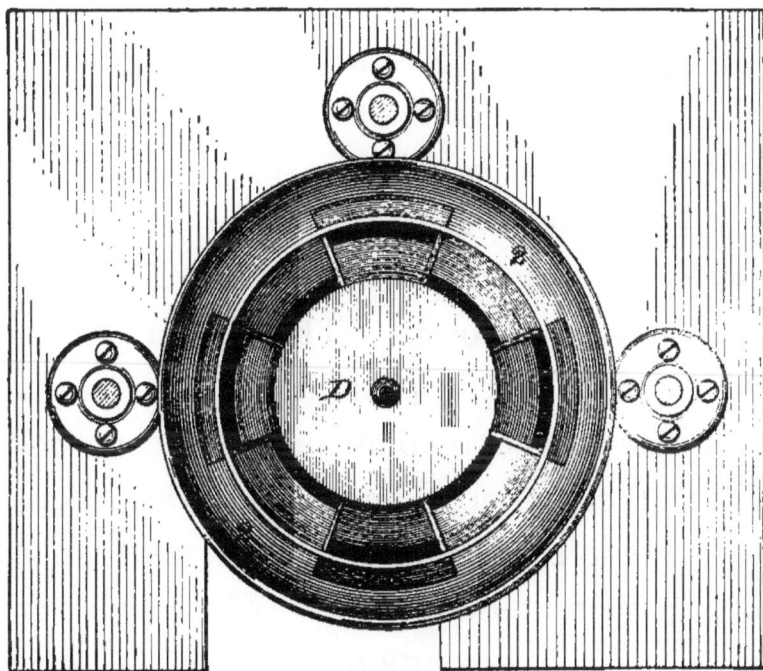

FIG. 12.

ticular field of usefulness is the quick and certain cutting down of the miscellaneous small samples, (from five pounds to five hundred pounds in weight,) that are constantly being received by all assay offices. It will handle anything from the finest assay pulp to crushed material of one-half inch or more in size. It is a very decided improvement over any of the present methods of quartering, or cutting down with sample shovel or tin sampler. The method of operating is as follows:

The material is fed either by hand, or (with large lots,) from a suitably supported bucket, into the funnel "F," the divider "D" being first set in rotation by hand, clock work or any convenient power. The divider gives, as will be seen by inspection of the drawing, eight cuts to the revolution, four being delivered to the funnel 1, and four to the receptacle 2; that is with uniform flow and speed, cutting the material in half. The divider may easily run 100 revolutions per minute, giving in that time 800 cuts, a very much greater distribution and division than can be secured in any other way. The rejected sample passes down the outlet to "O-2," the retained portion through the outlet "O-1," both into suitable vessels. The retained portion, should it be too large, may be cut again and again, until of suitable size. The operation is very accurate and very rapid, being about as fast as

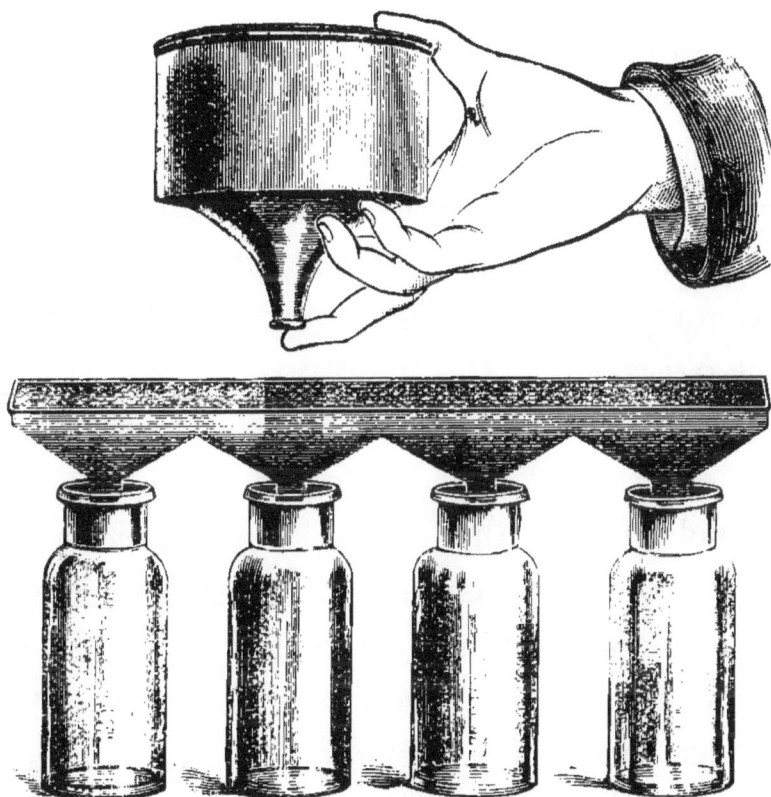

FIG. 13.

the material will flow through a one inch spout.

Ore Mixer and Divider.—Fig. 13 clearly shows this handy little piece of apparatus, also devised by Mr. H. L. Bridgman. We quote from the inventor:

"This apparatus entirely obviates the tedious and frequently inaccurate methods (usually with oil cloth and spatula) now in general use, for mixing and dividing the ground samples of ore, matte, slag and other similar material. An experience of several months has shown a very decided improvement in accuracy, speed and general convenience over the old way.

The operation is as follows: The ground material is introduced into the large, covered funnel, (mixer,) the outlet being first closed by thumb or finger as may be most convenient. Funnel and contents are then well shaken for a few minutes, and then, with opened outlet, passed to and fro over the set of distributing funnels (divider) and bottles as shown. With very finely ground or very light material the flow may be assisted by a slight shaking or tapping with the hand. The little skill necessary is readily acquired.

APPARATUS USED IN ASSAYING.

The mixer will also be found very useful for the prompt and thorough mixing of crucible assay charges, and all other work of similar character.

To test the efficiency of the mixer, a lot of 6 assay tons of litharge, 3 assay tons of soda, and $\frac{3}{4}$ assay ton of argols was taken, well shaken, divided by weight into three lots of

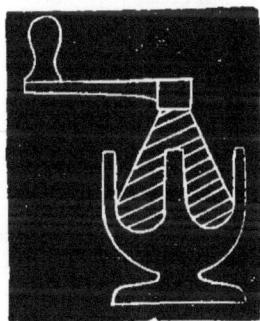

FIG. 14.

$3\frac{1}{4}$ assay tons each, and these charges fused separately in crucibles. The resulting lead buttons weighed 53,436, 35,416 and 53,398 grammes."

Fig. 14 illustrates "Buck's Patent Amalgam Mortar," or laboratory "arrastre." It may be used in place of the grinding plate.

Sieves.—A sieve of 80, 90 or 100 meshes to the linear inch is necessary. Such sieves are furnished of 5, 6, 7, 8, 9, 10, 12 and 15 inches diameter (6 to 8 inches is a good size), of copper or brass wire in a wooden frame. Those composed of horse hair are apt to deteriorate.

The box-sieves in brass are depicted in figs. 15 and 16.

Fig. 15.

Fig. 16.

A sieve which is better than the ordinary wood-bound pattern is one that is pressed or moulded out of the sheet gauze, and with no wooden rim to retain pieces of an assay which may afterward fall out to contaminate a succeeding assay. It would require but little of a rich ore to make a worthless one appear valuable by the above accident. Fig. 17 shows the idea.

Fig. 17.

If the system of sieving and sampling that

I speak of under the sampling of ores is adopted, there will be required a further set of four sieves of 2, 4, 8, and 16 meshes respectively, each 10 inches in diameter, and with frames of wood 3½ inches deep.

A 40-mesh sieve is useful for sieving certain chemicals, and two common flour sieves are wanted for bone-ash and granulated lead.

Zinc Sifting Pans. Fig. 18.—These will be

FIG. 18.

found convenient, and are better than paper for sifting over. A pair is necessary, the material sheet zinc. Length in full 31 inches, of body 25 inches, of neck 6 inches, width of body 12 inches, of neck 2 inches, height of rim 2¾ inches, with upper edge turned over heavy iron wire.

Spatulas and Spoons.— By a spatula we mean an instrument shaped somewhat like a

table-knife (fig. 19), and used for mixing
paints, ores, charges, etc.   It may be of iron,
steel, copper, platinum, silver, ivory, horn,
porcelain or glass. A large one of steel
or iron (such as painters use), length in full
$10\frac{1}{4}$ inches, blade $5\frac{1}{4}$ inches long by $1\frac{1}{8}$
wide, is a very good size for mixing ores
and crucible charges.   For weighing out ores

FIG. 19.

and mixing scorification charges, a smaller
one of steel, length 6 inches, blade $3\times\frac{1}{2}$ inch
is useful.

Two or three horn spoons, with or without
handles, are serviceable for various purposes.

SCALES AND BALANCES.

Two balances will be sufficient for the
ordinary work of the assayer; a small one
for weighing fluxes, ores, lead buttons, cupels,
etc., and a more delicate one for very accu-

rately weighing the gold and silver beads and gold residues.

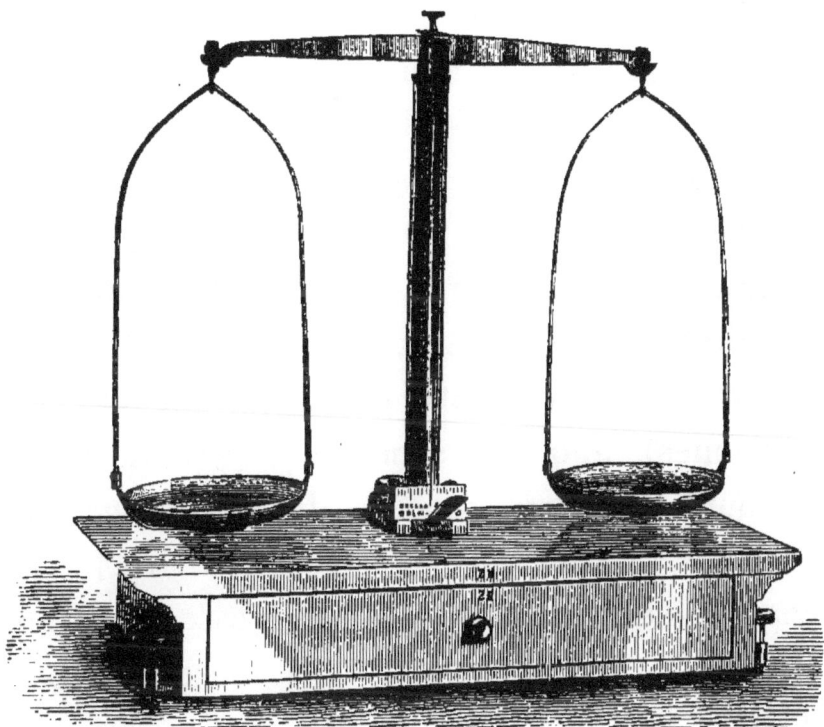

FIG. 20.

Scales for Pulps and Fluxes.—Considerable latitude can be allowed in the choice of such scales. Balances can be procured carrying 2, 5, 10, 20 or 30 ounces and upward, and ranging in delicacy from $\frac{1}{60}$ of a grain to 1 grain. For descriptions, illustrations, and prices of these and others, see the lists of the various manufacturers.

Not wishing to puzzle the student too much, I specify but four:

Fig. 20 represents as satisfactory a pair as can be wanted. It has a spirit level and two thumb-screws, movable scale-pans (3½ inch diameter), and when it is to be transported, all the parts on the box can be packed in the drawer. Its capacity is 10 oz. (about 311 grammes), and it is sensible to $\frac{1}{20}$ of a grain (about 3¼ milligrammes). Its cost is $22.00, (£4 10s.) (See Becker's list, No. 19.) A glass case in which to keep the scales is a good thing to have, and costs $6.00, (£1 5s.)

Troemner makes a similar balance, capacity 16 oz., sensibility $\frac{1}{20}$ grain, diameter of pans, 4 inches, price $18.00 (£3 14s.) (See Troemner's list, analytical scale No. 2, fig. 22.)

A still cheaper, but satisfactory balance, similar to the preceding, is likewise furnished by Troemner, capacity 8 oz., sensibility $\frac{1}{20}$ grain, diameter of pans 3¼ inches, price $15.00 (£3 2s.) It is represented in fig. 21. The objections to it are that it lacks the spirit level and adjusting screws, so that it is not always easy to get it into perfect equilibrium.

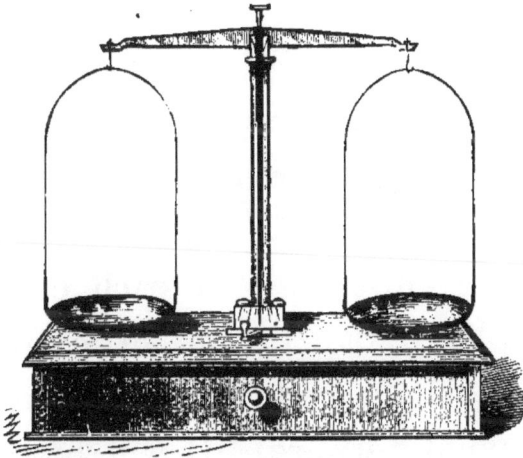

FIG. 21.

Directions for Setting-up and Testing.— The various parts of the scales shown as fig. 20 come wrapped and packed in the drawer beneath. They should be carefully unwrapped, rubbed a *little*, if necessary, with some soft, clean and dry buck or chamois skin, and put into place in the following order:

A brass wire, somewhat U-shaped, is to be

run up from the under side of the top of the box stand through two holes. The ends of this wire pass through two holes in the base of the pillar (which is now standing on the box), and are there held by two screws, shown in the cut. Next the swinging needle (also variously known as the· index needle, needle indicator, indicating rod, index pointer or pointer) is *firmly* screwed on to the centre of the beam, and the latter placed on the knife edges in the socket on top of the pillar.

Upon the knife edge at each end of the beam is placed one of the little frames or stirrups having a knife edge and hook, the latter to point forward. It will be observed that each end of the beam and each stirrup is marked with one or more dots. They must, therefore, be put together appropriately, that is, the stirrup with say two dots is to be placed on that end of the beam which also has two dots, and so on. The wire frames that support the scale pans are now taken, the brass piece at bottom of each swung out at right angles to the wires, and the frames suspended from the hooks at end of beam. The pans can then be placed in their proper positions. The leveling screws, shown at the front corners of the box, are now put in place, and the scale is ready foɪ testing and afterward for work.

To test, turn the leveling screws until the bubble of air in the spirit level is at rest in the centre of its circular case ; push down the lever shown in front of the gràduated scale, thus allowing the pans to swing freely, and gently vibrate the needle by directing a slight puff of wind from the hand upon either pan. The needle should vibrate to the same distance on either side, less a small fraction every time due to the decreasing momentum, and when it has finally come to a condition of perfect rest it should be exactly in front of the central division of the graduated ivory scale.

If the scales are not in good order there are no means of adjusting or correcting (save such as the natural mechanical ability of the owner may suggest), and the best plan would be to exchange them for a more perfect pair. This is not likely to be the case, however, more especially since such scales are not required to be very delicate.

The directions I have given above for the setting up of the scales shown as fig. 20 will apply, slightly modified, to any other and similar pairs.

Whatever pair be used, it is advisable to take two large watch glasses, in diameter a trifle less than that of the scale pans, and file down one or the other till they balance each other perfectly. Weigh all charges in these, thus avoiding any danger of corrosion or attrition of the scale pans.

A soft brush should be employed to brush out the contents of the glasses.

It is a good plan to *always* place the weights in one pan (the right-hand one), and whatever is to be weighed *always* in the other.

In case many crucible assays are to be made

(requiring the weighing of many fluxes), a great deal of wear and tear of the balances described can be avoided by substituting *hand-scales*, which are cheap and serviceable.

As it frequently becomes necessary to weigh larger quantities (ounces and pounds) of ores, fluxes, etc., it will prove a convenience to have at hand a reliable, quick-working balance, for

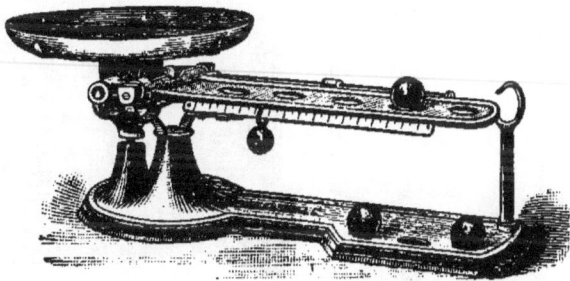

FIG. 22.

such work. Fig. 22 depicts a new form recently introduced by Troemner, and known as the "Ball Balance." It avoids the handling (and consequent errors) in the use of ordinary weights, and is capable of greater rapidity and

accuracy of work. The range is from ¼ ounce
to 16 pounds, and price is $14.00 (£2 17s. 6d.)

*Balances for Weighing Gold and Silver
Beads* —Here likewise is a range of choice,
and personal preference comes largely into
play.

When the assayer intends to travel consid-

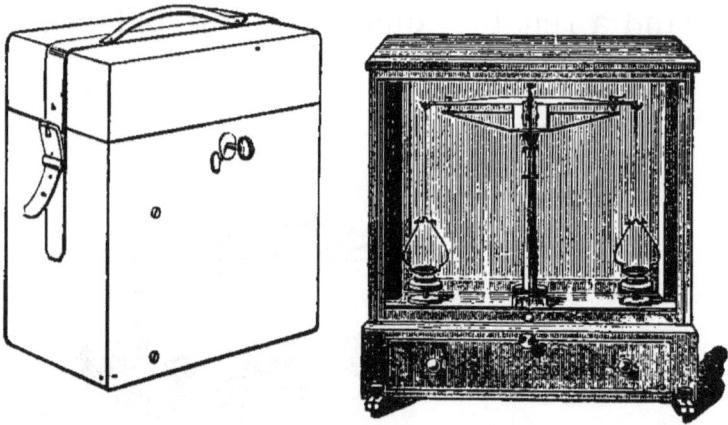

Fig. 23.

erably, the balance now spoken of will be the
most suitable.   Length 9 to 9½ inches, height
9¾ inches, width from 3 to 4 inches. It packs
into a light box, and by means of a strong
leathern strap can be carried by the hand.
Total weight, boxed, about 4½ pounds.   With
the proper weights this balance will weigh

$\frac{1}{10}$ milligramme, and by use of the swinging needle and ivory scale will indicate $\frac{1}{20}$ milligramme. In and out of case, it is illustrated in fig. 23.

It is made both by Becker (No. 2 of his list) and Troemner—the prices, with weights (1 g r a m m e down to $\frac{1}{10}$ milligramme), being $75 and $65 respectively. ($£15$ 8s. and $£13$ 7s.)

When the balance is not liable to be moved around

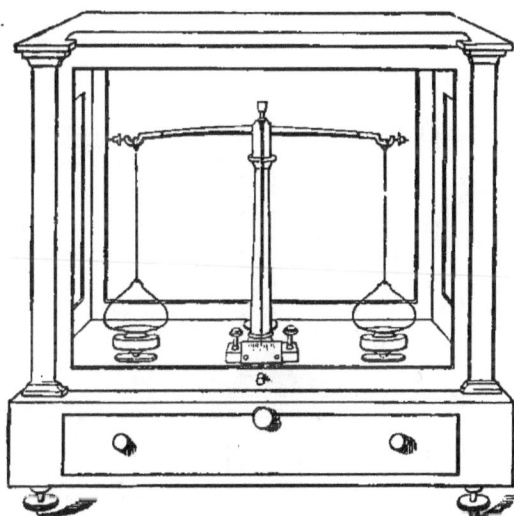

FIG. 24.

very often, that shown in fig 24 will serve nicely, and is considerably cheaper. The needle indicates $\frac{1}{5}$ of a milligramme, and each pan can bear a load of 25 grammes. Price $55.00, ($£11$ 6s.) of either Becker (No. 1) or Troemner (No. 1).

A similar but larger and somewhat more delicate balance is made by Becker (No. 3— price $78.00, £16). The needle indicates 10 divisions on the scale for 1 milligramme. The greatest objection to all of the preceding balances lies in the fact that they have no graduated beam to carry a rider to show the weight below 10 milligrammes, but instead indicate it by the deviation of a needle. The following pos-

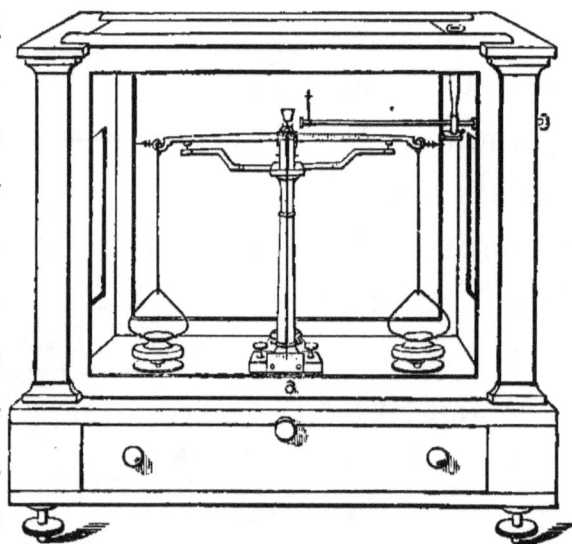

FIG. 25.

sess this advantage: Troemner's No. 2 (fig. 25 of this book) is a fine balance. The beam at the right is divided into *hundredths,* so that by means of the rider, a button of $\frac{1}{10}$ milligramme can easily be weighed. And more; since the spaces between the $\frac{1}{10}$ milligramme divisions are appreciable, by placing the rider half-way between any two, a weight $\frac{1}{20}$ of a milligramme can be determined. Price of this balance, $80.00 (£16 9s).

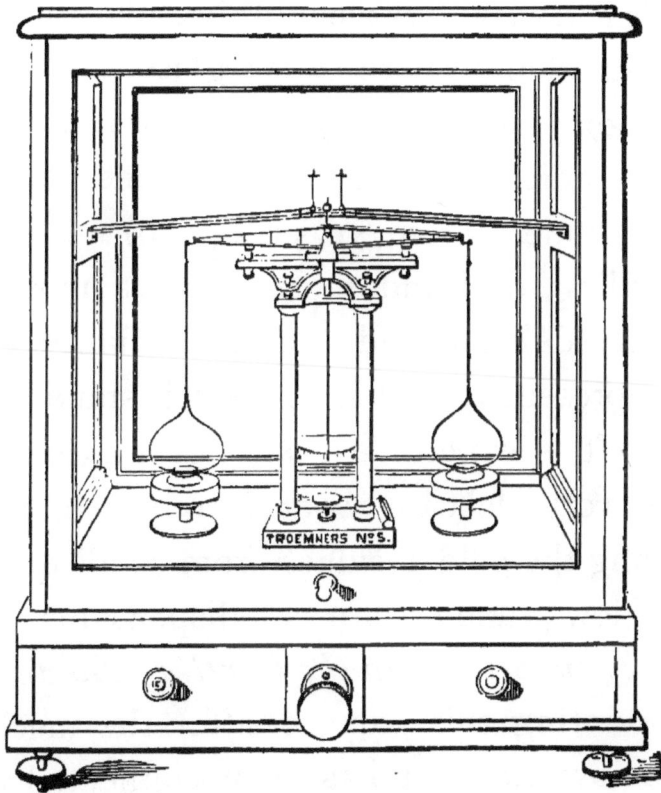

FIG. 26.

There is a still larger balance (Becker No. 5, Troemner No. 3—price $95.00, £19 10s. 6d.), which is more delicate and better finished than the preceding.

Fig. 26 is a good illustration of Troemner's "Extra Fine" new balance (No. 5), which is noticeable in having its beam of aluminium. Has a double column, with improved new eccentric lift, that works smoothly and regularly; beam divided on both ends; glass case large and roomy, with heavy plate-glass bottom; needle indicates forty full divisions for one milligramme. Price $175.00. (£36).

Oertling, of London, furnishes a most exquisitely delicate and accurate balance (No. 12), represented in fig. 27, which stands very high. It is expensive—$175.00 (£36)—but where very delicate work is required, as in weighing the gold residues from small quantities of low-grade ores, it is indispensable.

Notes on Setting-up the Oertling Balance.— In mounting this balance, after the mechanism below the floor has been connected with the beam supports, and the standards (or central pillars) have been screwed in position with the thumb screw provided, the beam (with its index pointer) is put in its place. This must be done from the top, and care should be taken that the pointer is not bent in so doing.

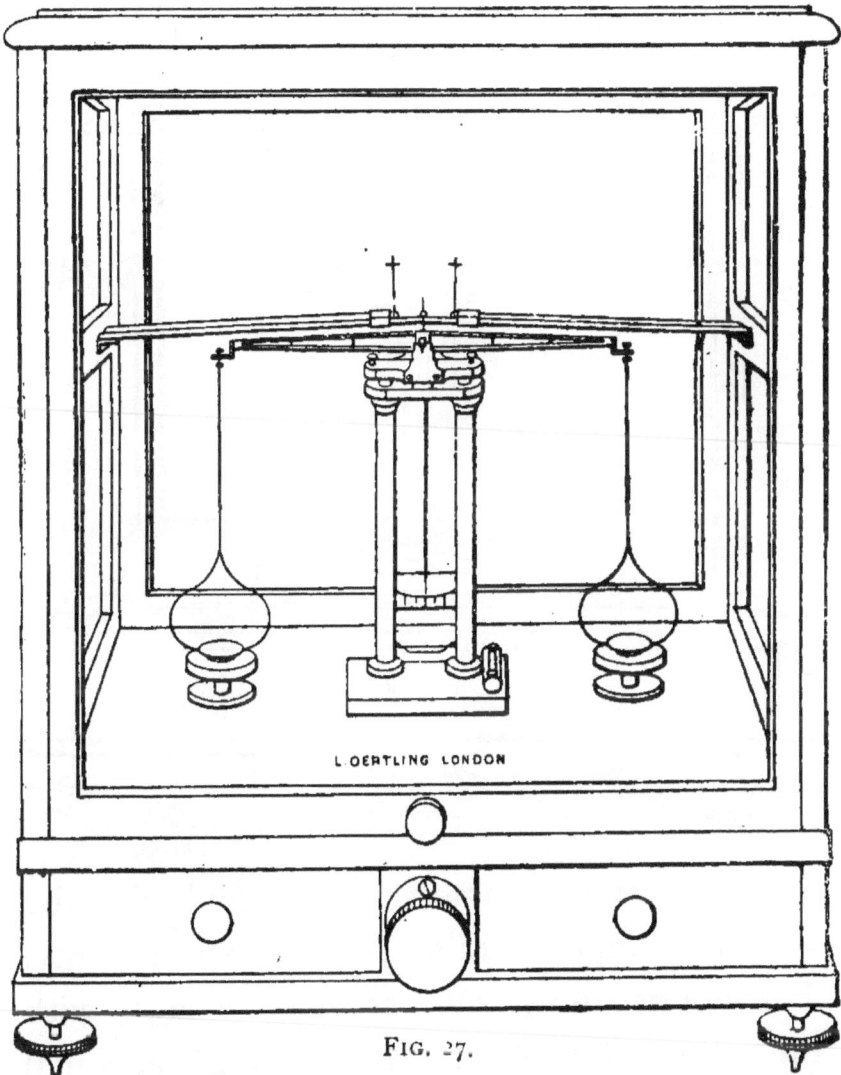

FIG. 27.

Now adjust the capstan-headed screws on top of the standards in such a manner that, when the beam is raised by turning the thumb screw in front, each screw will touch

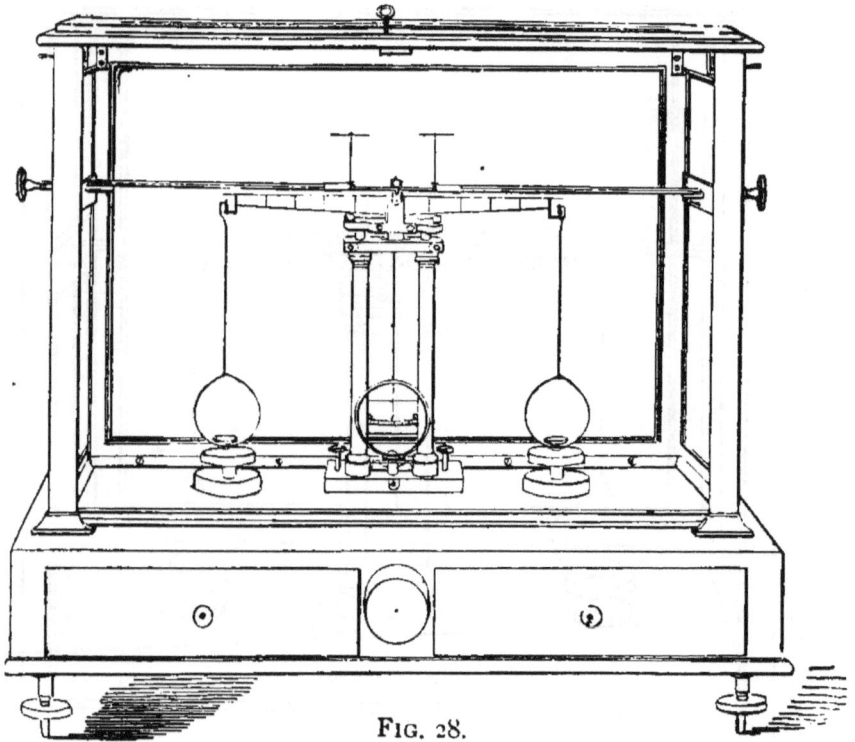

FIG. 28.

the beam at the same moment, and with the V-shaped support raise the beam uniformly. The pointer must also coincide exactly with the zero point of the ivory scale. A small pin is furnished with the balance for turning the capstan screws.

Now the pans may be placed in position and each pan support adjusted in height by means of its screw so as just to *touch* the pan, when the beam has been raised from its bearings.

Fig. 28 represents the Ainsworth balance, a recent but worthy rival to the Oertling. In general, it is similar to the latter, but is so constructed that any possible warping of the case will not affect the working of the balance. The rider attachment is so arranged that any lost motion can be taken up. A magnifying glass is adjusted in front of the needle to show the slightest deviation. Sensitive to the $\frac{1}{100}$ part of a milligramme. Price $175.00. (£36).

See the price lists of the manufacturers quoted.

It will be seen from an examination of the preceding illustrations that all the delicate assay balances are alike in their general con-

struction.   Each consists of certain characteristic parts, as herewith described.

First, a central pillar, or two pillars, firmly
attached to the floor of the enclosing case.
Upon the top of the pillar or pillars is a
plane bearing a V-shaped crotch lined with
polished steel or agate (preferably the latter,
in which rests the beam by means of a steel
or agate knife edge.   At each end of the
beam is fixed a steel or agate knife edge,
from which hangs a little frame with a steel
or agate plane, and from the frame is suspended, by a long and thin wire, the stirrup
carrying the detachable scale pan.

From the centre of the beam depends a
long, delicate index needle, which swings in
front of a graduated ivory scale.

From the right-hand side of the balance
case of the Becker and Troemner balances,
and from both sides of the Oertling and
Ainsworth, extends inward a movable rod,
controlled outside the case by a milled screw
head.   This rod manipulates a fine "clothes

pin " of wire known as a " rider," which can be placed on any point of the graduated beam.

By the employment of a simple mechanism the beam can be raised from its agate bearings to avoid unnecessary and wasteful friction or injury by sudden shocks. This is done by turning the milled screw head shown in front. The scale pans are supported when the beam is raised by two little disks coming up from below.

On the top of the index needle is a minute ball, which, by being either screwed up or down, raises or lowers the centre of gravity of the balance, and so either increases or diminishes its sensitiveness.

At each end of the beam of the Becker and Troemner balances is a very small milled screw head which can be screwed in or out, and by them the beam is balanced. On the Oertling and Ainsworth, instead of these screws, a fine piece of wire is twisted around the little ball on top of the beam, and points forward, and by turning it a little one way or the other the same object is accomplished.

A circular spirit level, or two tubular ones at right angles to one another, at the base of the central pillar or pillars, will indicate whether the balance is or is not level; the milled screw heads under the corners of the case are to do the regulating.

The whole mechanism of the balance is contained in a wooden frame with glass on the four sides. Both the front and back windows slide up and down, and are locked with the same screw or key which controls the beam.

Special Directions.—The workmanship of a fine balance is as delicate as that of a watch, consequently the greatest care should be used in setting it up and in handling its various parts. Many excellent balances have been ruined or greatly injured by pure carelessness, as by striking the knives, or by letting the beam fall suddenly on the central agate bearing, which will destroy the delicately ground knife edges, and, as a consequence, the high sensitiveness is lost.

The agate hangers should be placed on the beam so that the marks on each correspond — that is, each hanger or stirrup has a mark corresponding to one on the knife end of the beam to which it belongs, and, as the hangers are not made interchangeable, correct placing should be observed.

Care must also be used in handling the pointer or needle, that it is not bent. Should it scrape against the ivory scale when oscillating, lift off the beam, taking hold of it at the ends, and lay the pointer on a smooth, flat surface and gently bend the pointer downward; then replace the beam and note any improvement. If it still scrapes, remove beam again and repeat the operation, and continue this until the needle will oscillate perfectly freely. It is not advisable to have the pointer too far away from the ivory scale, as it makes it so much more difficult to read the result.

It must be remembered that *all* of the balances described are *very delicate pieces of apparatus*, and should be guarded with the *utmost* care.   They should be placed far from the heat of the furnace, and even away from the rays of the sun, tending to unequal expansion and subsequent contraction.   Shocks must be avoided and even *continued gentle agitation*, and they should be kept away from *acid fumes* (particularly those of *nitric acid*), and out of moist atmospheres.   By having a small vessel filled with *dry fused calcic chloride* always inside the case of the balance, the moisture present will be absorbed by it, and thus prevent, in a measure, the rusting of the steel parts of the balance.   When saturated replenish.

These balances are provided with steel knives and agate bearings, spirit level and set-screws.   By means of the latter and by observing the spirit level, the balance can be placed in a state of perfect equilibrium, *and it should always be kept in such*.   To ascertain whether it is in adjustment, throw the rest down, thus leaving the pans free, and vibrate the needle by a puff of wind from the hand. The needle should go to the same distance on either side, less a very small fraction due to the decreasing momentum. If it does, the balance is in equilibrium ; if it does not, adjust the difference by means of a little screw at one end or the other of the beam, or arrow in the centre.

To test its sensibility place a one-centigramme weight in each pan and a rider at equal distances on each side of the beam, and see if in balance. If not, adjust as before. Now move one of the riders one of the smallest divisions — the balance should respond quickly and distinctly, also with one-half of the smallest divisions (representing with a five-milligramme rider $\frac{1}{20}$ milligramme).

When not in use, the rest should not be left down, and on the other hand, when using the balance, it should not be brought up when the needle is vibrating, as this tends to throw the knives off the agate bearings, and so work injury.

With these balances, as with the ore scales, put the weights always in one pan, the material to be weighed in the other. Since the rider in most balances is used on the beam at the right, it is better to employ the right-hand scale pan for the weights. Do not leave the weights on the pan for too long a time.

In high, dry altitudes, use care in wiping the glass doors of the balance, as electrical action can be excited which may affect the accuracy of the weighing. Particularly should any rubbing be avoided just before using the balance.

Cover the balance when not in use with a thick woollen cloth, or heavy pasteboard box lined with flannel or similar material; this to keep out dust.

A very soft and fine camel's hair brush may be employed to cleanse the scale pans or other parts from dust.

Finally, never use these balances for any other purposes than for weighing gold and silver beads, small pieces of silver or gold, etc.

WEIGHTS.

The assayer needs three sets :

1st. A set in the French or metric system, ranging from 50 grammes down to 1 centigramme (10 milligrammes). Since they are not to be used for very accurate work, but only for weighing fluxes, lead buttons, and other comparatively rough purposes, they need not be very expensive — $5.00 to $6.00 (£1 to £1 5s.)

2d. A second set of metric weights, to be very accurate, their range from 1 gramme to $\frac{1}{10}$ milligramme, their use for weighing gold and silver beads. Such a set is included in the price given for the assay balance for gold and silver beads first mentioned. Separately, they will cost $8.00 to 10.00 (£1 13s. to £2 1s.) Fig. 29 represents a box of these weights.

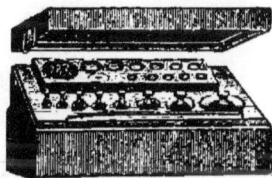

FIG. 29.

3d. A set of assay ton weights. These are important, and should be very accurate. Their range is from $\frac{1}{20}$ to 4 A. T., and price about $6.00. (£1 5s).

These weights are simply invaluable on account of their use requiring no calculations beyond a few multiplications or divisions.

But this assay ton system of weights seems to be one of the bug-bears surrounding the art of assaying to many beginners, especially those advanced in years and opposed to progress, and whose knowledge of weights is based entirely upon the arbitrary systems known as the troy, apothecaries', and avoirdupois. There is no difficulty in understanding it, and I think my explanations will make the matter clear to all.

The assay ton system is not restricted to any one system of weights — it can be applied to any, be it in use in whatever country.

First, then, to illustrate its use : An assayer weighs, of an ore to be tested, $\frac{1}{5}$ A. T. (assay ton). As a result he obtains a bead of silver weighing 10 milligrammes. If $\frac{1}{5}$ A. T. produces 10 mgrs. (milligrammes), 1 A. T. of the ore will produce $10 \times 5 = 50$ mgrs. of silver, and the assayer reports the ore as carrying 50 ounces of silver to every ton. If 4 A. T. of the ore were used and a 10 mgr. button obtained, then $\frac{10}{4}$ or $2\frac{1}{2}$ would be the number of ounces per ton of silver that the ore would produce. The simplicity of the arithmetic, rapidity of calculation, and difficulty of making mistakes are all apparent here; there remains only to explain the connection of the A. T. with our ounces, pounds, and tons.

In this particular case the French or metric system of weights is the one employed as a basis; but that is immaterial, as will be shown further along. To proceed: 1 ton avoirdupois = 2,000 pounds avoirdupois; 1 pound avoirdupois = 7,000 grains Troy; therefore, 1 ton avoirdupois = $7,000 \times 2,000 = 14,000,000$ grains Troy; 1 ounce Troy = 480 grains Troy;

hence 14,000,000 divided by 480 equals 29,166 ounces Troy in 1 ton avoirdupois.

Now the assay unit, called the assay ton, is (in this case) a weight of 29.166 grammes (a gramme being equal to 15.4 grains Troy), or, (1 gramme being equivalent to 1,000 milligrammes), 29,166 milligrammes. Hence the relation of milligrammes to ounces is as 1 to 1—in other words, a milligramme corresponds to an ounce, so that if by assay of 1 A. T. of the ore we obtain gold or silver to the amount of 4 mgrs., then, without any calculation, we know the ore will run 4 oz. to the ton.

The above calculation starts out with the short or American ton of 2,000 pounds. The long or British ton weighs 2,240 pounds, but we can use it in like manner as a factor, thus: $7,000 \times 2,240 = 15,680,000$; 15,680,000 divided by 480 equals 32,666.

Hence the unit of an assay ton system for Great Britain or Canada, based on the ton of 2,240 pounds, would weigh 32.666 grammes, and accordingly as we took fractions or multiples of it in assaying ores, so would our resultant beads of the precious metals be fractions or multiples of an ounce per the long ton.

Those who may object to the metric system can still use the assay ton system by making the unit a weight of 291.66 grains Troy, or 326.66 grains Troy. In actual work, it has been found that $\frac{1}{5}$ A. T. of the gramme system equals 5.83 grammes, equals 90.01 grains Troy, is a good quantity of ore to use in a scorification assay. An equivalent amount, or nearly so, in the grain assay ton system first given, would be $\frac{1}{3}$ A. T. equal to 97.22 grains.

The adoption of any system of assay ton weights avoids long calculations and the use of tables. By employing a whole number for the weight, then dividing the result by the said number and multiplying the quotient by 29.166 or 32.666, we obtain the same figures as by the assay ton system, but this necessitates considerable multiplication or the use of previously prepared tables.

Besides the above three sets, it will be found desirable to have a fourth set of grain weights, which may range between 1,000 grains and $\frac{1}{100}$ grain, or 300 grains and $\frac{1}{10}$ grain. They need not be extremely accurate. The first set will cost $10 or $11 (£2 1s. to £2 5s.); the second $2 (8s. 3d). It is sometimes necessary to weigh a bead, button or other object directly in grains, when these weights will be handy.

For large weighings in pounds, the proper weights will usually be found to accompany the scales.*

(See tables of weights in appendix.)

FURNACES.

There are *three* distinct and separate operations to be performed in an assay furnace—

---

* If many gold bullion assays are to be made, the assayer will find it extremely convenient to have a set of so-called "gold weights." (See the article "The Assay of Gold Bullion.")

roasting, crucible fusion, and muffle work (*i.e.* scorification and cupellation).

In an assay laboratory of any extent, where many assays are daily performed, it will be advantageous, if not imperative, to have a special furnace for each of the above classes of work, but ordinarily the assayer can manage to get along with one. It is requisite, then, that the one selected be adapted to carry on all the aforementioned operations. As to the particular kind he must consult his individual preference—and his purse. I cannot here describe all the varieties of furnaces which have from time to time been devised; all I can do is to speak of a few considered the best.

The heat-supplying medium of a furnace may be any one of three kinds of fuel, which fact, therefore, will serve to form a classification of the furnaces themselves into three divisions:

A. Furnaces employing gaseous fuel.

B. Furnaces employing liquid fuel.

C. Furnaces employing solid fuel.

Strictly speaking, the heat in any case comes from the combustion of a gas, for whether the fuel be liquid or solid, the burning matter is either the liquid transformed into gas, or it is gas driven off from the solid fuel. But the distinctions drawn will do well enough for my purpose.

A. Furnaces Employing Gaseous Fuel.

These are the so-called *gas furnaces*, meaning thereby that the source of heat is our common illuminating gas. But as this fuel will not be on hand for the majority of those for whom this book is written, they are respectfully invited to pass over the following dozen pages.

The furnaces I am about to describe are made by the Buffalo Dental Manufacturing Company. This company manufactures furnaces for either crucible or muffle work, or both, some of which require a blast, others only the natural pressure of the gas. The student is

referred to their circulars, also to Mitchell,
pages 79 to 100.

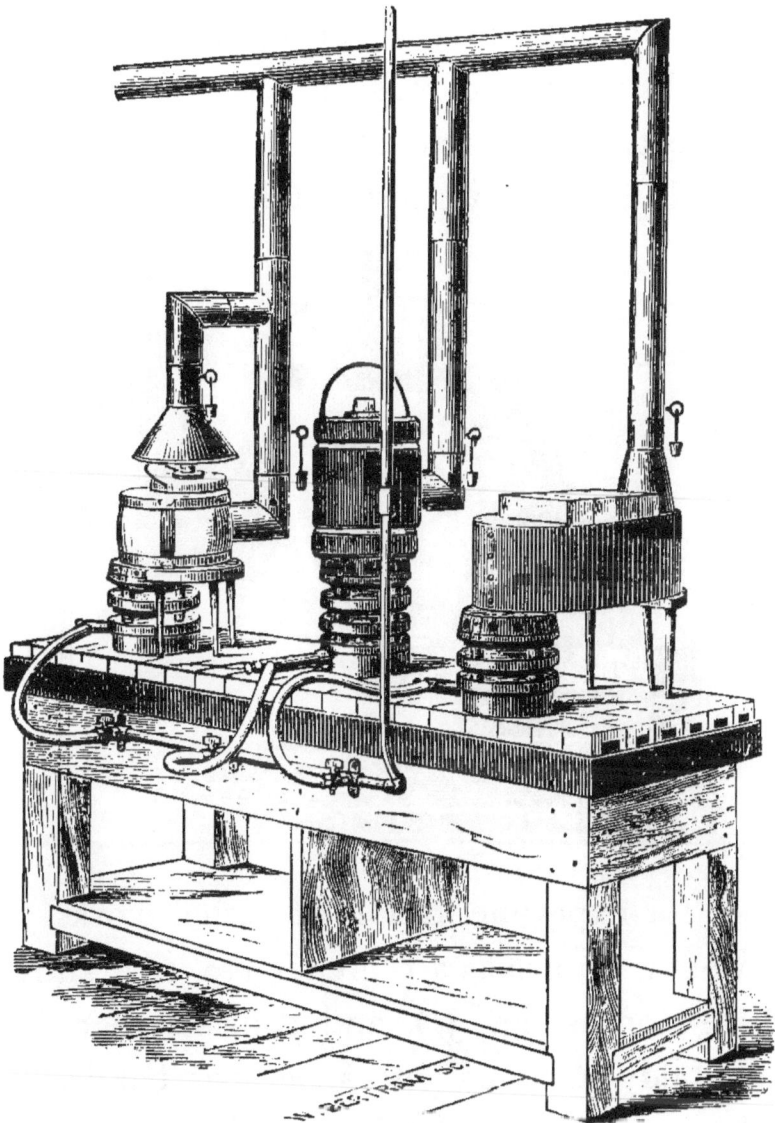

FIG. 30.

Figure 30 gives a very faithful representation of a group of gas furnaces as it was designed by and arranged for myself.

Its duty is to do roasting, crucible fusions, scorification and cupellation. The furnace at the left is for roasting sulphurets or other

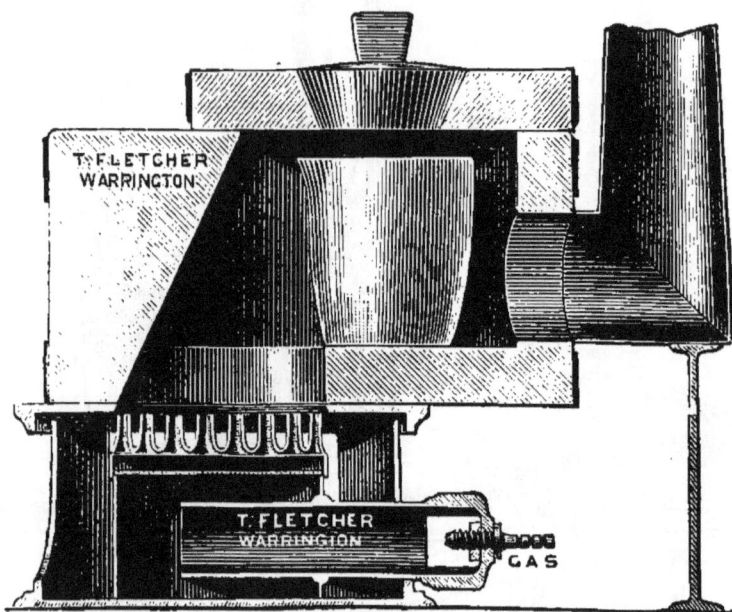

FIG. 31.

ores, for experimentation or actual work. It is what is known as a Fletcher No. 163 (shown in section as fig. 31), and consists of a fire-clay body strapped with sheet-iron bands, and a burner (No. 16 Fletcher). The opening at the top (protected when not in use by the cover shown) is to allow the heat to have full play upon the roasting-dish placed on it. The heat and flame pass from the burner through the furnace and out and up the chimney-pipe. The funnel-shaped pipe over the cover is to catch and draw the fumes up the chimney. When the burner is lighted a most powerful draft ensues, carrying all odors and fumes at once away. Both the pipe and hood are provided with dampers, controlled by small weights. The burner is connected to the gas-tap by stout $\frac{1}{2}$ inch bore rubber tubing. A cast-iron tripod supports one end of the furnace and keeps everything firm.

Next in regular order (supposing a sulphuret ore to be under treatment) is the middle furnace, for crucible fusions. It, likewise, consists of a furnace and burner. The latter is a Fletcher No. 15, of same construction, however, as the No. 16. The furnace proper (shown in detail in fig. 32), is made in five parts, the central section (a cylinder of

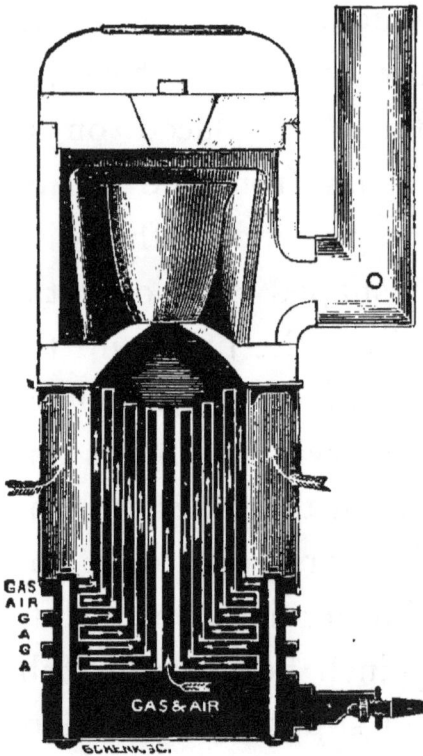

GAS
AIR
GAS
GAS
GAS

GAS & AIR

SCHENK SC.

FIG. 32.

fire clay), the bed-plate upon which it rests, and which has an opening for the flame to pass through, the cover (with handle attached), and which also has an opening filled by a plug, all of fire clay, and finally a plumbago lining, rubber tube, chimney connection and damper, as with the other furnace.

Finally, at the right, is shown a furnace for scorification and cupellation, and which I have, I think fitly, designated as the "Monitor." Fig. 33 shows it enlarged and uncovered. In form it is almost that of the reverberatory furnace, the movable bricks, when in place, being the roof. Looking at it from another point of view, it may be described as a muffle with the flame as well as the heat inside. Its exterior dimensions are as follows: 20 inches long, 7 inches wide and $5\frac{1}{2}$ inches deep. In the interior, upon the bot

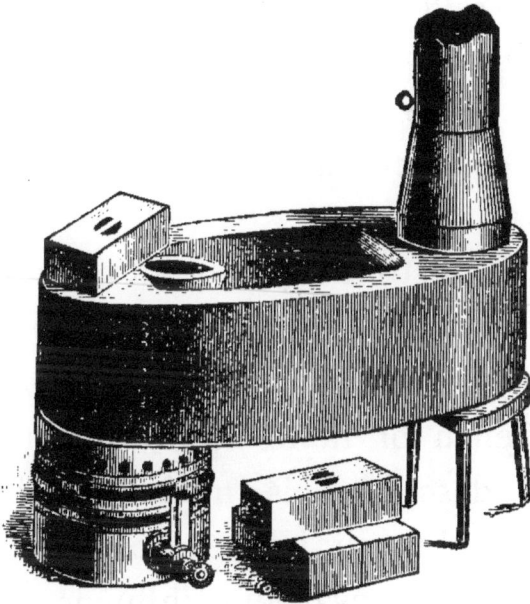

FIG. 33.

tom, are four little wedge-shaped bridges of fire clay, which are movable, and upon them rests a false bottom or floor, also movable. The latter corresponds to the muffle bottom of an ordinary furnace, and upon it is done all the work. It is $3\frac{1}{2}$ inches wide by $7\frac{1}{2}$ inches long and $\frac{1}{2}$ inch thick, and has a shoulder or bench running across its entire width on the end nearest the burner. The covering bricks, four in number, are each 7 inches long by $2\frac{3}{4}$ inches wide and $1\frac{3}{4}$ inches high, each with a slotted bridge for its convenient handling. The burner is the No. 16, Fletcher. Similar connections to the first-mentioned furnaces.

The 3-inch stove pipes of all three furnaces are fitted into one long horizontal pipe, which fits snugly into the chimney.

The bench or table upon which rest the furnaces described, is made of pine, well seasoned and firmly jointed, to resist as much as possible the warping influence of heat and to support the weight of the furnaces and table tiles. Its dimensions, not figuring on the top, which overlaps 1 inch all around, are: 4 feet 6 inches long, 1 foot 7 inches wide, and 2 feet 1 inch high plus the thickness of the top, which is $1\frac{3}{4}$ inches. A double coat of shellac varnish is its sole ornamentation. To the sides and ends of the table top are firmly screwed four strips of band iron of $2\frac{1}{2}$ inches width and $\frac{1}{8}$ inch thickness, and of such lengths as to alternately overlap at the angles, making smooth joints. The top of this sort of wall is $\frac{3}{4}$ inch above the bed of the table. Upon the latter are 114 fire-clay tiles, or, rather, clamps, such as are used to join house tiles, and having the shape of the letter " E " less the middle projection. Their average size is $3\frac{1}{2}$ inches by 3 inches across and $1\frac{3}{4}$ inches high. They are so arranged on the table as to leave a series of six air tubes or chambers running its entire length. The spaces between the tiles are filled with a mixture of plaster of Paris and Venetian red rubbed up with water. The latter color is also used for the tiles themselves, and somewhat on the fire-clay portions of the furnaces.

A $\frac{3}{8}$-inch bore gas pipe, with proper taps and nozzles, is screwed to the front of the table.

I have been somewhat lengthy in the above detailed descriptions, but have done so for the benefit of such as may care to duplicate the outfit.

The manner of operating the furnaces is simple. As regards the roasting furnace it is merely to shut off the dampers of the other furnaces, turn on and light the gas, and regulate the heat to suit the particular ore. (See under *Roasting.*) The control of the mixture of gas and air is made by means of the milled handle at the burner.

Next as to the crucible furnace. Remove cover, turn gas on full at tap, light, and regulate by milled handle. Crucibles that contain charges that are to be heated gradually can be placed in the furnace as soon as lighted; others after the lapse of a few moments, to allow the furnace to become thoroughly heated. Placing the charges in cold, I have made good fusions of refractory ores in 25 minutes from time of lighting. The furnace will take crucibles in size up to the Battersea

"S," which is 4¼ inches across by 5 inches deep.   Use no covers.

Finally the "Monitor," for which I may claim, not originality (that belonging to Mr. Thomas Fletcher), but merely applicability and decided improvements in the form, the original one being the roasting furnace already described.

To manage it, remove the covering bricks, open the damper and shut those of the other furnaces, turn full flow of gas on, light as usual, and replace bricks.   In from 15 to 30 minutes the interior will be hot enough. Remove one or two bricks nearest burner, place charged scorifier on false floor, and replace bricks again.   When the charge is melted, slide aside brick nearest burner, and set one of the floor supports diagonally into the furnace, one end resting on bridge of floor, the other will protrude above the top of the furnace.   This is to *break the flame*, and is *absolutely necessary* in order to introduce air for oxidation.   In cupellation, the gas is turned down more than in scorification.

The time of performing either scorification or cupellation (which see) varies according to the nature of the ore, charge, size of button, etc., but is about the same as that occupied in the use of a coke furnace.

The consumption of gas is not far from 30 cubic feet per hour. It is not intended nor claimed that this furnace can take the place of one required to be run from 10 to 12 hours per day, for here, of course, a solid fuel will be cheaper, but for small runnings of from 1 to 4 hours or so it is economical, as are also the others.

For small laboratories, then, the advantages of this furnace are many: convenience of operating, whereby the assayer sees every step and stage of the operation, and so can tell when and where to change or improve ; comfort in manipulation, for it does not heat up the vicinity of the furnace and the room itself (quite a desideratum in the summer time);

perfect control of the source of heat, so that a higher or lower temperature, a reducing or oxidizing effect may be produced in an instant; entire noiselessness, in which characteristic it is the superior of all blast assay furnaces; saving of time, which, for furnaces employing coke, charcoal or coal, is spent in "bedding down," feeding, breaking coke, etc. freedom from the annoyances of dust, ashes and smoke; absence of waste; and, finally, its remaining qualifications, which need not be dwelt upon, are simplicity of construction, durability and portability.

The complete plant, as illustrated, costs very close to $75.00. (£15 8s.)

B. Furnaces Employing Liquid Fuel.

There are many varieties of furnaces that come under this heading, their fuel being refined petroleum, gasoline, etc. As with gas furnaces proper, some are intended for crucible fusions, others for muffle work ; still others are for both. The air pressure in some forms is derived directly from foot bellows, in others from air compressed by an air pump.

(Consult the circulars of the Buffalo company already mentioned for descriptions of their furnaces ; also Mitchell, pp. 72 to 78.)

Fig. 34 represents Hoskins' Hydro-carbon Assay Furnace.

Fig. 34.

The apparatus shown in full above has now been in practical operation for several years in many parts of the country, and is past the experimental stage. There are many inconveniences and annoyances necessarily connected with the use of coke and coal furnaces whereas this apparatus does away with all dust, ashes, constant replenishing of fuel and a large amount of radiated heat; in fact it has all the advantages of a blast gas furnace, with the additional advantage that it may be forced to almost any extent without the use of a blower, and being at the same time automatic. The maximum expense of running is about five cents per hour, and in our large cities will not exceed three cents. Although there has been and is a prejudice against gasoline as a fuel, this apparatus, as will be seen from its construction, is perfectly safe, and no accident can occur except through gross carelessness.

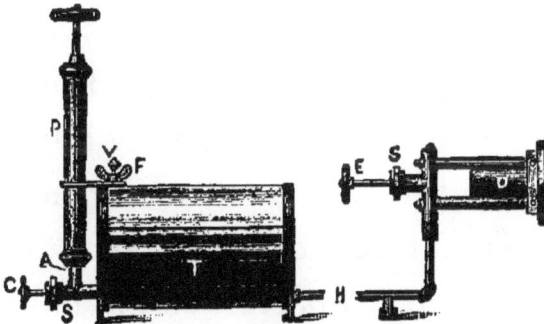

FIG. 35.

Fig. 35, P is an ordinary force pump, at the bottom of which, at A, is a valve which closes automatically upon releasing the pressure from the pump; C is a check valve, which closes the inlet to the tank T completely; F is the filling screw; V is the vent screw, for letting off the pressure when through; H is the pipe leading from the tank to the burner D; E is the burner regulator, terminating in a fine point, closing the orifice of the burner; S S are packing boxes. Upon opening C and pumping a few strokes a pressure is created in the tank and on top of the fluid, which is forced through the tubes of the heated burner, vaporizing the gasoline, which finally issues from the orifice at the end of E as a highly heated gas, and burns as such in a powerful blast. After once starting, the heat of the flame passing through the burner vaporizes the fluid in the tubes, and hence the apparatus is automatic; requires pumping up only once every quarter of an hour or half hour, according to the power of blast desired. Its action may be controlled from the heat of an ordinary Bunsen burner to that required to melt cast iron, etc.

Fig 36 represents the muffle furnace (three sizes, taking Battersea C, F and L respectively), fig. 37 the crucible furnace (two sizes),

FIG. 36.

FIG. 37.

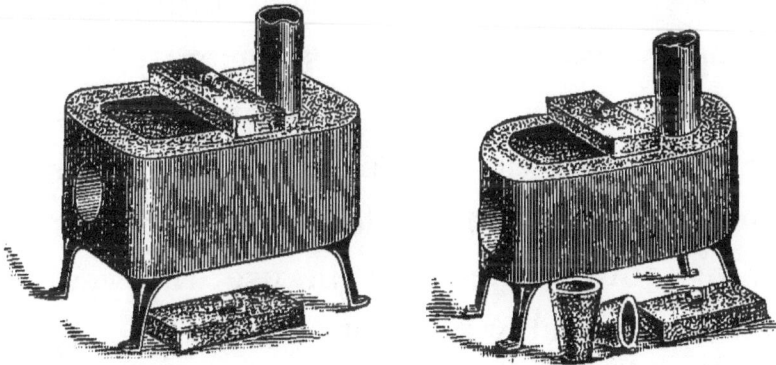

FIG. 38.

while fig. 38 represents another form of crucible furnace (also in two sizes).

Directions For Use.—Close E, unscrew F, and introduce about two quarts of gasoline, (of 74° Beaumé,) according to the capacity of the tank. Replace F and close V ; open C one or two turns, and give three or four strokes of pump P, and close C. Heat the burner by igniting some of the fluid in a suitable vessel placed under the burner (an old scorifier will do to hold the gasoline) ; when hot enough apply a match and open E gradually until the action is more or less uniform. If no spray or liquid issues from the orifice, the burner is hot enough. If not hot enough, burn slowly until no liquid or spray issues. When sufficiently heated the blast can be made of any intensity desired by the use of the pump as above. To stop its action, shut the regulator E, or open screw V, or both ; when not in use the vent V should invariably be kept open. The mouth of the burner D should be two to three inches from the fire hole of the furnace.

For high temperature and muffle work proceed as follows :

First, light as above, and heat inside of furnace to redness at least.

Second, place the burner against the inlet of furnace.

Third, turn out the blast with E and immediately turn it on again without lighting it (or simply blow the flame out of the burner tube), when, if the furnace is hot enough, the gas will ignite inside of the furnace. The heat can be regulated as in the first method of burning. When burning *inside* of furnace there *must be no flame in the burner tube.*

C. Furnaces Employing Solid Fuel.— In this class are found the best known furnaces, using as fuel, wood, charcoal, coke, hard and soft coal, or mixtures of them. The furnaces themselves may be divided into two general classes, portable and permanent, accordingly as they may be intended to be moved about or remain fixed.

I. PORTABLE FURNACES.

Very many forms have been manufactured. Space compels me to reduce the description to three ; the Battersea, Bosworth and Brown's.

The *Battersea* furnace is of fire clay, made
in sections and bound with iron bands.　Fig.

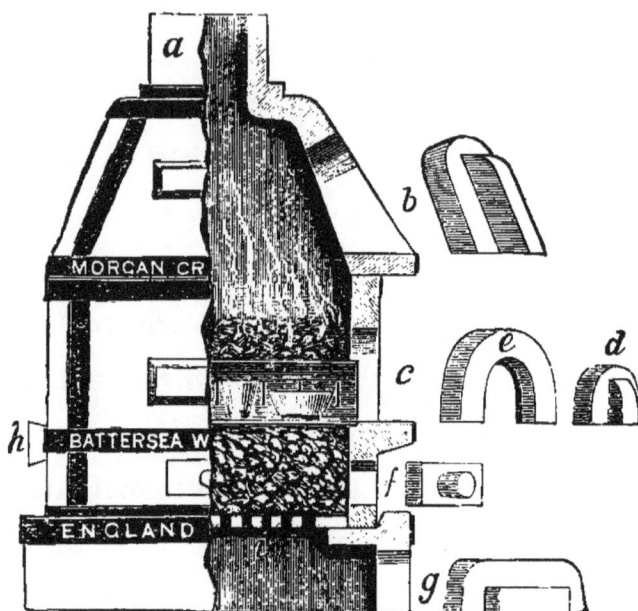

FIG. 39.

39 is a sectional view of the muffle furnace, of
which the following sizes are obtainable in
this country :

No.	Height, in.	Diameter, inches.	Size of Mufflers, in inches.			Price.			
			Long.	High.	Wide.		£	s.	d.
C	27	14½	9	3⅝	5½	$25 00	5	3	
D	28½	15¼	10	4	6	30 00	6	12	6
E	29½	16¼	12	4	6	35 00	7	4	
F	30	17½	14	5	8	40 00	8	4	6
K	48	23	15	6	9	80 00	16	9	

The Bosworth (Fig. 40) is also of clay, in three sections, securely bound. Its construction is such that it is not liable to crack. It is durable, convenient, does much work on little

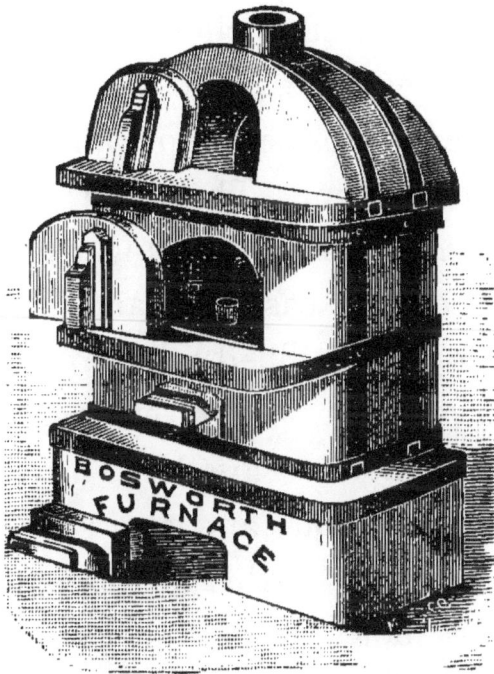

FIG. 40.

fuel, and heats the muffle quickly and evenly, It uses a 9x15 muffle, although it can be made to use a 10x16, if desired. Price $40.00 (£8 4s 6d.)

For details concerning the preceding furnaces consult the various catalogues and circulars of the manufacturers.

Brown's Portable Assay Furnace. Fig. 41. —This furnace consists of a nearly square sheet-iron frame 28 inches high, 14 inches deep, 16 inches wide, lined with fire brick in sections, the interior being smooth and straight from top to bottom. The cover is cast-iron, and is *ridged* to lessen the danger of cracking. The muffle door is cast-iron, and is fitted with a circular opening, filled with *mica*, that the operations going on within the muffle may be seen when the door is closed. The draft-doors are also of cast-iron, and are provided with wheel openings to further regulate the draft. The circular holes at bottom are in all four sides of the furnace, and serve to keep cool the true bottom of the furnace upon which the ashes fall. The corners of the castings are rounded to prevent breaking.

The muffle seen in the opening rests equally

upon the fire-brick in front and in the rear of the furnace, leaving a space of $\frac{1}{2}$ inch be-

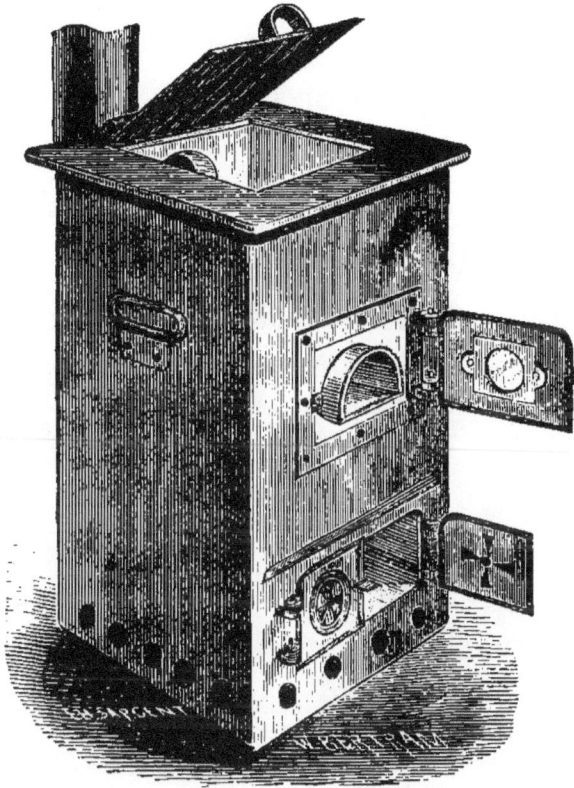

FIG. 41.

tween the end of the muffle and the brick to allow the passage of fumes. There is also a space for fuel of 4 inches on each side of muffle.

The grate is formed of cast-iron bars, 10 inches long, 1 inch wide, 9 in number, resting upon a cast-iron frame.

The space below the true bottom is to be filled with fire-brick or sand or other material convenient.

The chimney hole is 5 inches in diameter, thus accommodating a stove-pipe of same dimensions. The bottom of this hole is 17 inches from the true bottom of the furnace, and 8 inches from the bottom level of the muffle.

There is a handle upon each side of the furnace to allow more convenient handling.

The furnace will take a J Battersea muffle, 12 inches long, 6 inches wide, and 4 inches high. But a muffle, of any of the various brands found in the market, of dimensions approximating those given, can be used. Entire weight of furnace is 155 pounds.

The above furnace possesses the following advantages:

1st. Simplicity. Having no complicated parts to get out of order.

2d. Usefulness. It can be used both for muffle work and for crucible operations.

3d. Capacity. There is no other furnace manufactured of similar dimensions and weight which can accommodate so large muffles, and consequently produce so much work and so rapidly.

4th. Durability. Being made of heavy sheet-iron, it cannot be broken by handling nor injured by heating.

5th. Adaptability. Any fuel may be employed for which the draft of the chimney is sufficient.

6th. Light weight. This furnace weighs but 155 pounds, as against 300 to 400 pounds of other furnaces.

7th. Cheapness. It is from one-third to two-thirds cheaper than any other furnace that will do as good work. Boxed for transportation, $20.00 (£4 10s. 6d.)

(See advertisements in the appendix.)

II. PERMANENT FURNACES.

Whenever an assayer becomes permanently settled in any locality, it may pay him to erect a brick furnace, which, under such circumstances, possesses some advantages over the so-called portable furnaces.

For descriptions of these, whether to be used for roasting, fusion, scorification and cupellation work, see Mitchell, pp. 57, 63, etc. The various stamp mills, smelting and sampling works, and mining corporations scattered throughout the West, have usually permanent furnaces burning coke, charcoal, soft or hard coal, which may profitably be imitated.

Fig. 42 gives a vertical section, and fig 43 a ground plan of a good furnace, the front view or elevation of which is shown in the frontispiece. The illustrations being to scale ($\frac{1}{32}$) need but little explanation. The muffle half of the furnace is at the right of each figure, the crucible furnace at the left, both constructed of fire brick.

The lower half of the muffle furnace is anchored with $1 \times \frac{1}{2}$ inch wrought iron bars and ties, and the upper part with those $1 \times \frac{3}{8}$ inch.

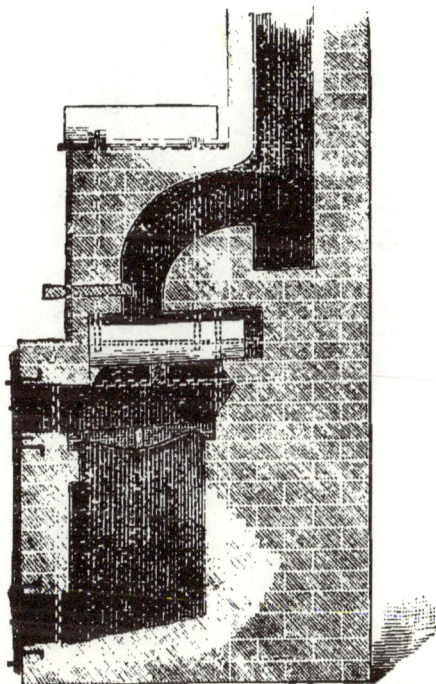

FIG. 42.

The muffle appropriate for this furnace is the L Battersea, $15 \times 9 \times 5\frac{3}{4}$ inches. Coke or coal may be the fuel.

The above very convenient modification of Plattner's furnace was constructed for Mr.

John C. Jackson, Metropolitan Block, Chicago,
and although erected for his especial work, is
yet suited to the necessities of almost any
assayer.

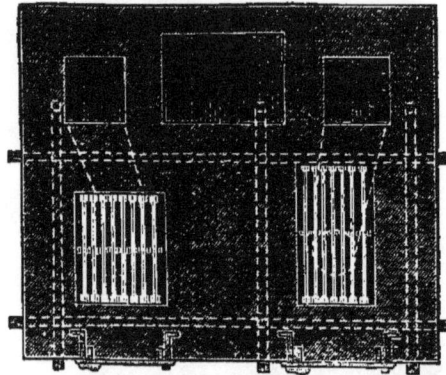

FIG. 43.

Furthermore, many a peripatetic assayer
has conjured up a temporary furnace of clay,
adobe, or home-made bricks, using a tile or
large crucible for a muffle, to meet an emergen-
cy, and when its days of usefulness were over,
left it to decay and ruin.   On such occasions,
necessity is indeed the mother of invention.

FURNACE TOOLS.

Crucible Tongs.— For placing in, and re-

moving from, the furnace, crucibles.   The assayer will need one'large and strong pair, of
wrought iron, 24 to 36 inches long, the grip-

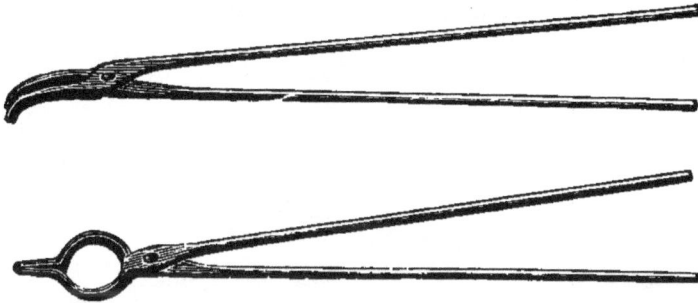

FIGS. 44 and 45.

ping ends of which may be curved or straight,
like figs. 44 and 45, or tongs can be procured

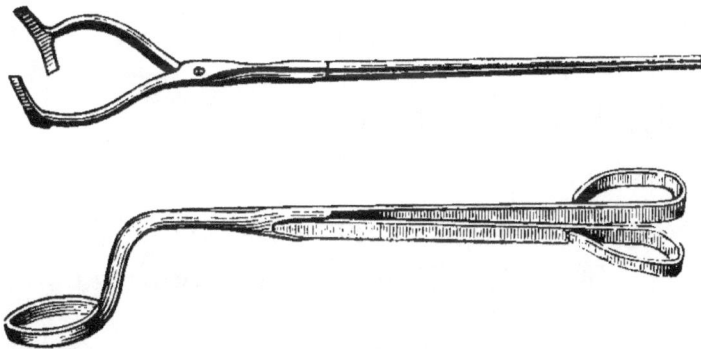

FIGS. 46 and 47.

of either one of the two forms given in figs.
46 and 47.   Either one, however, of the two
first mentioned, will do very well.

A smaller pair, from 15 to 18 inches in length, for lifting small crucibles and large scorifiers, for placing lead buttons in the cupels, and for opening and shutting the doors of the furnace, is invaluable (fig. 48).

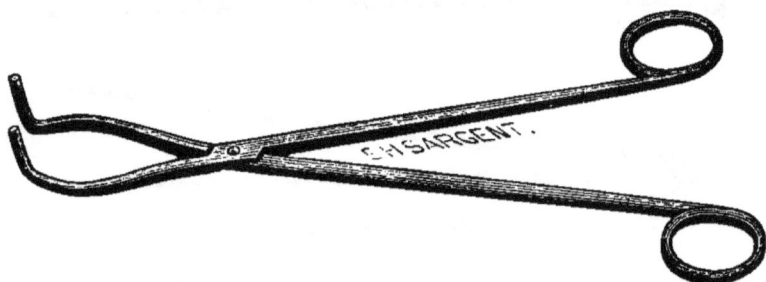

FIG. 48.

A still smaller pair of 8 inches in length (fig. 49), for managing the doors of the furnace, is handy, but not absolutely necessary.

FIG. 49.

Scorifier Tongs.—The correct shape is here given (fig. 50). The length about 24 inches.

FIG. 50.

The curved arms fit the bottom of the scorifier, the long arm extending across the top.

The best material for these tongs is steel, and they should not possess too much spring. Two or three sizes should be procured, to accommodate the various sizes of scorifiers.

Fig. 51 represents Judson's patent steel scorification tongs, arranged for lifting scorifiers from the rear of a muffle without disturbing those in front.

Fig. 52 shows a special and peculiar form of scorifier tongs required if the "Monitor" gas furnace described on pp. 76–78 is employed, or, for that matter, whenever a scorifier is to be raised by a perpendicular lift. Length about 15 inches, greatest width, 3 inches, and can be made of quarter-inch wire.

FIG. 51.

FIG. 52.

Cupel Tongs.—Several forms are permissible. A common pattern is shown in fig. 53.

FIG. 53.

In using these particular tongs, care should be taken to secure a firm grasp of the cupel, lest it slip through the tongs and be broken, and the bead be lost. On the other hand, too much pressure may crush the cupel.

With these tongs the grip should be made nearer the top than the bottom of the cupel, for should the operator happen to grasp with it a cupel *below* the latter's centre of gravity (especially when the cupel is soaked with litharge), it will probably turn over, again giving a chance for loss of the bead of precious metals.

FIG. 54.

The form next figured (fig. 54) is not likely to cause the above accident, nor quite

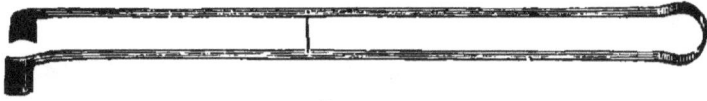

FIG. 55.

so liable to crush the cupel, and fig. 55 illustrates a better pair than either of
the preceding.

Fig. 56 represents Judson's
patent cupel tongs.

Whatever pair of tongs is used
should be of steel or wrought iron,
light weight with not too strong
a spring, length from 18 to 24
inches, and with a strong guide.

For the " Monitor" gas furnace
previously mentioned, and as a
companion tool to the tongs illustrated on page 101 (fig. 52), the
cupel tongs herewith shown as
fig. 57 is given. The arm "a" is
straight, "b" is curved to form of
cupel. Length about 18 inches,
material quarter-inch wire.

For considerable of my work

FIG. 56.

FIG. 57.

I have used, instead of tongs, the cupel shovel (fig. 58), and cupel rake (fig. 59). The curve of the latter fits the cupel. By means of these two implements, one or two cupels can be easily and quickly run in or out of the muffle without danger of damage. They can be of light weight wrought iron, and about 24 inches long. For carrying a half dozen cupels at a time, a second shovel with the blade six to eight inches long would be serviceable.

FIG. 58.

All the tongs, etc., just described (with the exception, perhaps, of Judson's, which are patented), can be made by any blacksmith, or even by the assayer himself by exercising a little home talent.

FIG. 59.

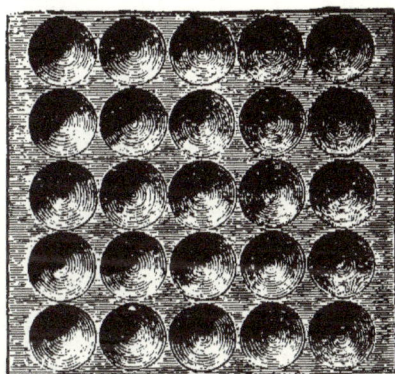

FIG. 60. FIG. 61.

FIG. 62.

FIG. 63.

Scorifier or Scorification Moulds, Slag Moulds, or Pouring Plates.—Four forms are here represented. Any hardware merchant can provide a very good substitute for any of the above

in the shape of a so-called "*gem-plate*," generally with twelve cavities.

Similar moulds can be procured of heavy
copper, which has the advantage of not suddenly chilling the slag and thus causing it to
retain small pellets of the lead, but they are
quite expensive, and if the iron plates are
warmed before using, they will serve quite
as well.

I do not recommend the practice of painting the interiors of the cups with ruddle or
chalk washes.

The utility of scorification moulds is obvi-
ous; by employing them the time of cooling
is greatly diminished, and the scorifiers can
(but they had better not) be re-used.

Large moulds for receiving fused crucible
charges can be procured (for example a
plumber's lead pot), but as mentioned elsewhere, unless crucibles are rare, they are
not necessary. It is the custom in some
assay offices to pour a crucible fusion into
one of the cavities of the scorification mould

with conical cups (fig. 62). The lead button sinks to the bottom of the mould and
the excess of slag, fused salt, etc., runs over
the top and into any convenient receptacle.

Muffl Scraper.—Shown in fig. 64. Made

FIG. 64.

of wrought iron, length 24 inches. The cupel
shovel spoken of can be used for conveying
sand or bone ash into the muffle whenever
lead has been spilled upon its floor, and the
scraper employed for bringing out the pasty
mass formed, and smoothing down the surface of the floor of the muffle.

Pokers.—One long straight one, 32–36
inches, of ⅜ inch wrought iron (fig. 65), a

FIG. 65.

short one, 18 inches, with end bent, for
muffle work, and a third ordinary poker, for
stirring the fire, are desirable.

APPARATUS USED IN THE FURNACE.

Muffles.—The term muffle is applied to that piece of apparatus figured here in various forms, in which are performed the operations of roasting, scorification, cupellation, etc. Muffles are made of iron, plumbago, or a refractory mixture (*i.e.* sand and fire-clay), most generally the latter. They can be procured in the market in almost every conceivable size (the price lists enumerating some fifty), besides which they can be made to order of any special dimensions.

Figs. 66 and 67 show the shapes most in favor; the latter being especially designed to accommodate the "Colorado" crucibles side by side. Still other muffles are open entirely at both ends, so that their contents can be manipulated as well from the back as from the front. Any muffle can be easily con-

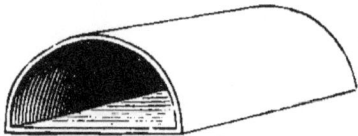

FIG. 66.

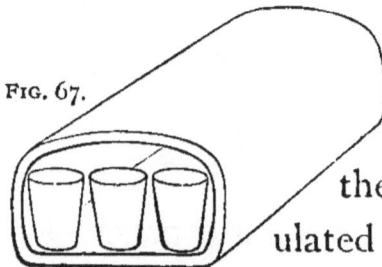

FIG. 67.

verted into this form by sawing off the closed end.

The size of the muffle employed will be determined by the size and make of the furnace.

In case the furnace is so constructed that the muffle can easily be taken out (and it is well to have it out during the firing up and first heating), then to avoid danger of cracking it by sudden heat it is best to place it on top of the furnace to warm it somewhat before putting it in position.

False floors to the muffle are obtainable, and save the real floor from injury due to spilled lead, etc. They are made of same material as the muffles.

Crucibles.—These vessels are made of various materials : black lead (graphite or plumbago), French clay, Hessian sand, charcoal-lined (*i.e.* Hessian sand crucibles with a lining of charcoal and molasses), quick lime, a mixture of magnesia and chloride of magnesium, alumina, and finally those for very special purposes, of porcelain, iron, platinum, gold or silver. It is but few of these varieties that are needed for the assaying of gold, silver, copper and lead ores, and these I now specify.

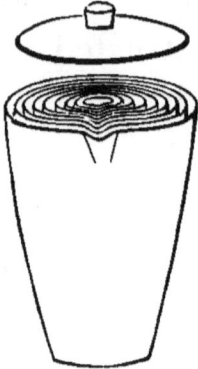

FIG. 68.

FIG. 69.

FIG. 70.

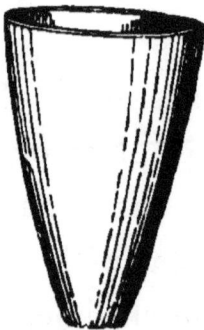

FIG. 71.

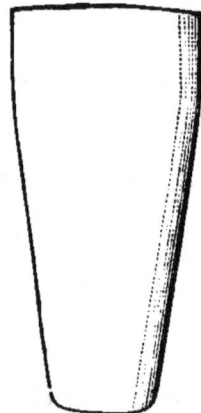

FIG. 72.

The so-called clay or sand crucibles are the ones fitted for the assaying of the ores of the four metals named. They occur in two forms, round and triangular (figs. 68 and 69), with covers to match. Almost any size can be obtained from some one of the manufacturers. In giving the charges for crucible work I have generally indicated the sizes of crucibles needed. The most commonly used crucibles will range between $3\frac{1}{4}$ and $4\frac{1}{2}$ inches in

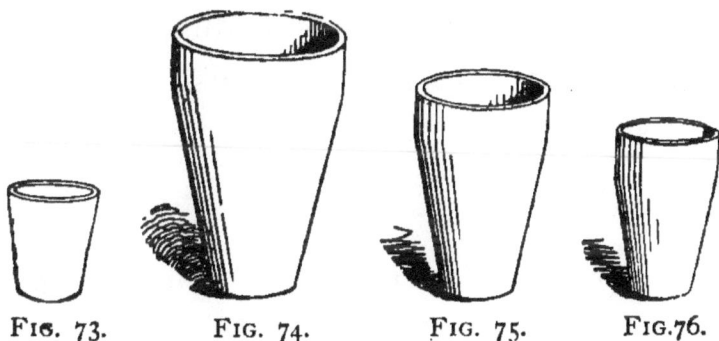

FIG. 73.   FIG. 74.   FIG. 75.   FIG.76.

height, and between $2\frac{7}{8}$ and $4\frac{1}{8}$ inches across. For an ordinary crucible charge of one assay ton and fluxes, a crucible $4\frac{1}{2}$ inches high by $4\frac{1}{8}$ across will be about right.

Fig. 70 shows the form of the French clay or Beaufay crucible or "fluxing pot."

Figs. 71 and 72 outline the forms of the Denver Fire Clay Co.'s crucibles.

For muffle lead the specially shaped crucibles figured here (figs. 73-76) are desirable. The largest size can also be used for fusions of gold and silver ores in muffle. They are made in three sizes. Those furnished by the Battersea company are known as the "Colorado" crucibles. Those manufactured by the Denver Fire Clay Co. are known as 5, 10 and 20 "gramme," according to the size as shown in the following list, dimensions approximate :

AA, "Colorado," or "5 gramme," 2¼ inches high; 2¼ inches across.

A, "Colorado," or "10 gramme," 3 inches high; 2¾ inches across.

B, "Colorado," or "20 gramme," 3½ inches high; 3 inches across.

Roasting Dishes.—These are made of refractory clay or black lead, of the form indi-

FIG. 77.

cated (fig. 77). They should be quite shallow. They are used for the roasting of ores containing much antimony, arsenic, sulphur and zinc. They are furnished in sizes ranging from 2 to 8 inches in diameter. The 3-inch dish is suitable for roasting $\frac{1}{5}$ A. T. The 5 and 6 inch sizes can be employed with satisfaction for open air roastings of 1, 2 or more A. T. (for crucible work), in place of the frying-pan.

Scorifiers.—These articles are made of a material similar to that of the clay roasting dishes. Fig. 78 shows the right shape. They should be somewhat shallow; in texture uniform, and free from cracks and holes.

FIG. 78.

They can be procured in sizes varying from 1 inch to 5 inches in diameter.

The best size for all ordinary scorifications is the $2\frac{3}{4}$ inch (if the muffle is wide enough to admit it). This size takes from $\frac{1}{5}$ to $\frac{1}{2}$ A. T. of ore, according to its gravity. The $2\frac{1}{4}$ inch is well adapted for re-scorifications, that is, for reducing in size too large lead buttons. It can also be employed when very little of the ore is to be worked, say $\frac{1}{10}$ A. T.

Have the $2\frac{3}{4}$ inch in large quantity, with one half as many $2\frac{1}{4}$ inch and perhaps a few

2½ inch. For certain other purposes it is advisable to have on hand a few of the 3 and 3½ inch sizes.

As the manufactured scorifiers will stand a great deal of rough handling without injury, and since they are well made and cheap, it is better to purchase them, rather than to attempt their home manufacture which is not a very easy thing.

Cupels.—Among the most useful articles the assayer possesses. They are employed to absorb oxides of almost all the metals save those of gold and silver, thus leaving these two metals behind in a state of comparative purity. Lead is the metal whose oxide, litharge, they absorb in great quantity. Any substance which will absorb these various oxides would do, but for many reasons, burnt bones or bone-ash is preferred. Good bone-ash is so easily and cheaply obtained that it seems a waste of time to more than indicate the process whereby the assayer himself may make his own supply. In brief, horse or sheep bones are boiled repeatedly in water, their organic matter (grease, carbon, etc.) burnt away, they are then finely ground, sifted and washed. (Mitchell, pp. 133-4.)

Very good cupels can be purchased in several sizes, and *when* they are good, can be safely packed and transported.

The one chief objection to purchased cupels is their expense, therefore ordinarily it is cheaper to make them, to do which I now give directions:

The bone-ash which can be obtained in bulk and of several grades, is mixed, say one pound at a time, with a strong solution of pearl-ash (or carbonate of potash) in warm water, till the mixture adheres well together, though it must not be at all pasty. (The right degree of moisture is hard to describe but easy to acquire.) When a portion of the mixture is squeezed in the hand, it should cake together (somewhat like half-melted snow) and show the imprint of the fingers. Now sift through a common flour sieve, place the cupel ring upon a block of wood (having a large piece of brown paper spread out below all), fill about flush with the surface with the sifted bone-ash and strike the plunger into the ring four or five times moderately heavily.   Turn the plunger around in the ring once or twice and push the cupel gently out.   A little practice will soon enable the assayer to turn out perfect cupels.

The moisture remaining in the cupels can be driven out by placing them on the top of the furnace after a day's running, or, what is better, by allowing them to dry in the normal atmosphere of the room or by exposure to the sunlight.   Cupels thus slowly dried are less likely to crack on using.

The texture of the cupel, that is, its degree of porosity, depending on the fineness of the bone-ash and amount of compression, is quite important.   If too fine bone-ash is used, the cupel will crack (or "check," as it is sometimes termed), in the muffle; if too coarse, the cupel will absorb silver, causing loss.

Therefore a medium grade had best be chosen. The above two difficulties are in a measure obviated by making the body of the cupel, that is, the lower two-thirds, of coarse material, and the upper third of fine.

If the cupel is too compact, cupellation proceeds too slowly; if too loose or porous, the cupellation proceeds too rapidly, causing a certain absorption of silver with the lead. As in everything else, experience is the best teacher.

The *form* of the cupel is immaterial. Fig.

FIG. 79.

79 represents the one which I prefer on account of the ease with which it can be removed from the mould.

A cupel with diameter of 1½ inch is a convenient size.

Annealing Cups.—Shown in fig. 80. Used

FIG. 80.

in the assay of gold bullion. Should be well made, light but strong. Various sizes can be obtained.

Annealing Plate.—Employed for annealing a number of slips at once, in the gold bullion assay. In size about 6 inches long, 2 wide and ¾ inch thick.   May be made of an old muffle floor rubbed down.   Can be purchased of either fire clay or plumbago.

APPARATUS OF GLASS AND PORCELAIN.

FIG. 81.

Sample Bottles.—A number of these (fig. 81), of two, four, six and eight ounces capacity, with wide mouths, and cork stoppers, are desirable for pulverized samples of ores.

Re-agent Bottles.—The dry re-agents are best kept in wide-mouthed bottles (known as "salt mouths"), glass stoppered, thus preventing the admission of dust and moisture.

FIG. 82.

Wet re-agents in bottles of kind illustrated in fig. 82.

Stone-ware crocks of various sizes can be employed instead of the bottles, and will, of course, contain greater quantities.   Fruit-jars with threaded necks and metallic caps will stand transportation better than the bottles, and tin cans or wooden boxes will pack more closely and last longer than either. Circumstances will alter cases if the laboratory is to be more of a traveling than a fixed one.

FIG. 83.

Whatever receptacles are used should be properly labelled.

Bottles, of course, are neccessary for the wet re-agents. The distilled water can be preserved in *clean* demijohns enclosed in wicker-work, or in clean stone jugs.

Wash Bottle. Fig. 83.—To contain distilled or pure water. A quart is the best size. By blowing in at the opening *a*, a fine stream of water is thrown out through *b*.

FIG. 84.

Watch Glasses. Fig. 84.—More correctly known as clock-glasses. A pair is desirable to place in the scale pans of the ore scales, to keep injurious substances away from contact; in diameter they should be slightly less than that of the pans.

Porcelain Capsules or Crucibles.—For holding the bead of gold and silver while being parted. Two sizes are convenient, one being 1 inch in diameter across top by ¾ inch in depth, the other 1½ inch diameter by 1$\frac{1}{16}$ deep. A good shape is that here figured. A dozen of each size will last some time.

FIG 85.

Test Tubes.—Used in qualitative tests. It is well to have some of four, six and eight inches in length. A rack to hold them is convenient. Some assayers employ them for parting gold and silver beads.

Parting Flasks (or Boiling Flasks). —A small flask, capacity ½ ounce, of form as figured, is sometimes used for parting gold and silver beads, instead of test-tubes or porcelain capsules. At least

FIG. 86.

three will be needed. Round-bottomed flasks are also frequently used.

Matrasses.—Flasks of the shape delineated in fig. 87 and used for the parting of gold bullion, are generally termed matrasses. It is quite important for the purposes of manipulation that the neck

FIG. 87. of the flask should fit snugly into the annealing cups employed. Can be held by a wooden clamp.

Besides the form figured, there are others used by assayers in foreign countries. Thus one variety is about the same as that in fig. 87, but lacking the shoulder. A third pattern is similar, but has a lip for pouring. A fourth has a very broad base and so on. Whichever form works best in practice with any assayer will, naturally, be preferred by him.

Evaporating Dishes.—Are of glass, and also of porcelain, the latter being much more durable. Used for evaporating liquids to dry-ness, or in place of the casserole figured on page 123.

FIG. 88.

Graduated Apparatus. — An accurately graduated burette (see fig. 88) of 50 c.c. capacity, and several glass-stoppered flasks of 1 litre capacity (1,000 c.c.), 500 c.c., 250 c.c., 100 c.c., 50 c.c., etc., will be needed for volumetric work. They can be purchased sufficiently accurate for all ordinary demands. Some measuring cylinders (lipped), of 10, 15, 25, and 50 c.c. capacity, not carefully graduated, are very useful in measuring solutions for the various analyses.

Glass Beakers. — Will be needed in the copper and other analyses, chlorination and other tests, etc. They should be lipped (fig. 89), and preferably of thin material to stand heat. Several nests may be wanted.

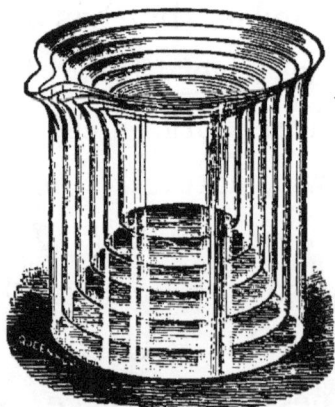

FIG. 89.

Glass Funnels.—For analyses of different kinds. Should be of an angle of 60° (fig. 90.)

FIG. 90.

Glass Stirring Rods.—Very useful. Cut up a long glass rod into various lengths, and round each end by holding in a lamp or gas flame for a minute or so.

Flasks.—One will be wanted for the chlorination assay of gold. Several sizes can be made use of, for wash-bottles, to retain solutions for any length of time, etc. Should be of thin glass.

Separatory Funnel.—See "Chlorination Assay for Gold," in appendix.

Casserole. Fig. 91.—Of porcelain. Can be put to many uses, as small evaporations, etc.

FIG. 91.

Pipettes. Fig. 92.—A 10 c. c. and a 5 c. c.

FIG. 92.

will be required in the copper analysis. They

can be home-made by drawing down to a fine opening one end of a glass tube, and rounding the other.

FIG. 93.

Mortars and Pestles.—Small sizes of these are useful in pulverizing re-agents, etc. Their material may be either glass (fig. 93) or porcelain ; shape as represented.

MISCELLANEOUS APPARATUS.

Note-books.—Indispensable. Nothing should be left to the memory, but everything important relating to the assay of an ore should be down in black and white.

The number of the ore, its character, the charge for the furnace, conduct in the fire, results of the various operations, as shown by the crucibles, scorifiers, cupels, slags, buttons, beads, etc., and all calculations, should be taken note of.

Gummed Labels.—An assortment of various shapes and sizes will be found extremely convenient.

Boxes.—Of pasteboard, 5½ inches long, 3½ wide, and 2 high, to be used for pulverized samples. Paper boxes, tin boxes, paper bags, and cloth bags are also used.

Paper.—Sheets of heavy brown or manilla paper for the mixing of ore samples are necessary.

Sheets of black glazed paper can be used instead, but it is better to reserve these for the mixing of charges, as they are a little too delicate for rough work. Some assayers use pieces of sheet rubber, rubber cloth or oil-cloth.

Whatever kind be employed, see that it has no holes to allow loss of sample or charge.

Tissue paper for wrapping up borax glass into pellets, and for enfolding minute gold and silver beads for flattening, will be needed. Filter-paper is indispensable for filtrations; also valuable for removing small quantities of moisture from the interiors of the porcelain capsules in the operation of parting.

Clean blotting-paper will do for the latter purpose.

The filter paper may be obtained in sheets or cut round of any size wanted.

Brushes.—Several are necessary. First, in case the rubbing-plate is employed, a large brush, such as is used by painters, is invaluable.

For brushing charges from the scale-pans or glazed paper, a medium size camel's hair is wanted, and for brushing the scale-pans of the delicate balances a very fine camel's hair brush is needed.

Pincers.—A few pairs of varying sizes are handy. One of about 8 inches in length, strongly made of wrought iron, a 4-inch pair of brass, and a third pair with limbs running down to a fine point, for picking up minute gold and silver beads, will suffice.

Hammers.—While the assayer can get along with one or two hammers, it is better to be provided with four or five. A heavy 5-pound sledge-hammer, a couple of smaller ones of about two pounds, one of them having one sharp edge and a square face (fig. 94), the other with both faces blunt, a small hammer for breaking crucibles and scorifiers and flattening buttons, and a ½-ounce sharp-edged hammer for trimming small specimens and flattening gold and silver beads, are very convenient.

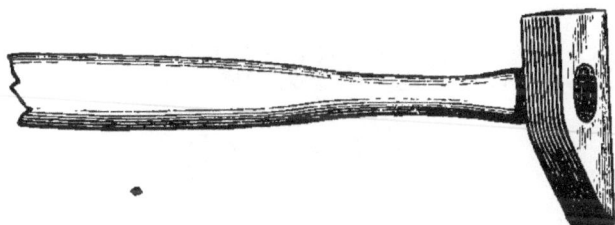

FIG. 94.

A sharp hatchet for kindling-wood and a dull one for breaking coke complete the category.

Anvils. — A miniature blacksmith's anvil (fig. 95), weighing 10 pounds, and properly mounted on a block, will be in constant demand. A simple yet satisfactory method of mounting the anvil has been of long-time use in my laboratory. An oaken block, 30 inches high by 12 inches through in both the other directions, has a frame of 1 inch wood screwed to its sides at the top, rising one inch above the surface. On the top of the block is nailed a half inch thickness . of rubber belting, leaving one-half inch space between its surface and the upper edge of the surrounding frame. The anvil is screwed

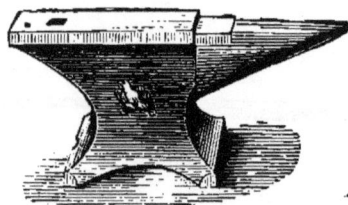

FIG. 95.

down to a piece of 2-inch oak fitting this space. The oaken block furnishes a firm support for the anvil, the rubber deadens the sound of blows, and by lifting off and putting aside the anvil and its bottom, the block serves as a convenient table for breaking ores in the mortars. An illustration of this is shown on p. 21 (fig. 3).

Another useful form of block and anvil

may be easily constructed. Obtain a good section of a tree trunk, such as butchers use, to be about 36 inches high and any convenient width, as 20 inches. In the centre excavate a hole somewhat smaller than the horn of the anvil figured above, and with a wooden mallet or block of wood and a heavy sledge hammer drive the anvil firmly home.

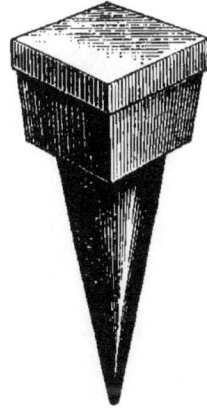

FIG. 96.

A flat plate of steel, $1\frac{5}{8}$ by $1\frac{1}{4}$ inches and $\frac{1}{4}$ inch thick, on which are to be flattened the gold and silver beads, is useful.

Ring-stand.—This implement, made of cast-iron, is useful for many purposes; to hold a wire triangle that supports the porcelain capsule used in parting, to support a sand-bath, wire-gauze, etc. Fig. 97 shows one pattern.

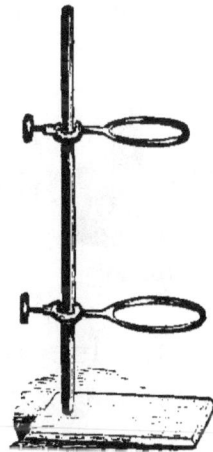

FIG. 97.

Wire Triangle. — Of twisted

FIG. 98.

wire (best of platinum), in shape as figured, for supporting capsules, etc. It may also be strung through pieces of pipe stems.

Sand-baths.—Any flat plates of tin or iron filled with sand. Their use is to distribute the heat around any vessels imbedded in the sand.

Wire-gauze. — Three-inch squares of iron wire gauze are used for same purposes as the sand-baths.

Burners, Lamps, and Stoves.—When gas

FIG. 99.

can be procured, the Bunsen burner (fig. 99) is the best supplier of heat for small purposes. By turning the ring at the bottom so as to close the holes, a light-giving flame is produced ; by leaving the holes open, there is obtained a heating flame due to the more perfect combustion. A large alcohol lamp is the best substitute for the Bunsen burner.

Lamps for gas, gasoline, kerosene, etc., are numerous. One of the best, the "Kellogg Bunsen Vapor Lamp," is shown in Fig. 100.

Note following directions: Fill reservoir ⅔ with 74° gasoline. Open valve at bottom, take off cap, blow in top to force air out of pipe. Turn wheel H until fluid escapes at E, let run until F is half full,

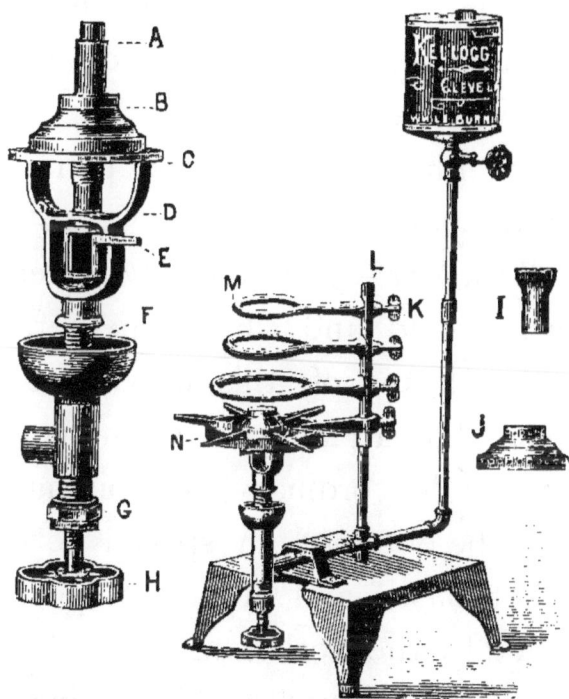

FIG. 100.

shut off wheel and close D. Light cup F and allow it to burn nearly out. Open D, turn H slightly, then light at A and C. Consult minor directions.

Frying Pan.—Aside from any culinary importance, this kitchen utensil serves a useful end in receiving melted borax glass, spreading the latter out that it may cool in a thin sheet.

It is also occasionally employed in the roasting of sulphurets, etc., on a comparatively large scale.

In either case coat the pan with chalk or ruddle paint.

Blowpipe.—For testing minerals and for fusing gold and silver together. There are many forms of this important little instrument, but a plain curved one is as satisfactory as any for ordinary blow-piping. They can also be obtained in special forms to give a "hot blast," and with a support for the lips, whereby the worker may not be easily tired. (Consult the works on Blowpipe Analysis.)

Cupel Moulds.—For making cupels. These are made of either steel or brass, but preferably of the latter, since they do not rust so quickly. A mould generally consists of three parts, the plunger or pestle, which is convex at the bottom to form the concavity of the cupel, the ring into which the plunger partly or wholly slips, and a bottom plate upon which the ring rests. In some moulds this bottom plate is circular and fits into the ring.

Fig. 101 represents a good form, which is of brass, and is furnished in sizes that make cupels of 1¼ and 1½ inches diameter. It has no bottom plate, but a smooth block of hard wood will serve equally well. The cupel this mould furnishes has its sides at right angles to the base (see fig. 79). One advantage this form of mould possesses is that by using more or less bone-ash, cupels of varying thicknesses can be obtained by reason of the plunger sliding *in* the ring, which is not the case with all others.

FIG. 101.

A special machine has been devised for making cupels, but I am not at all certain that it turns them out any better than does the common mould, nor more rapidly. If greater pressure is needed than that given by the hammer or mallet, a second-hand letter press might be utilized by knocking off the upper plate and making a few alterations.

Shears.—For cutting gold and silver bullion, sheet silver, lead-foil, etc.   Should be

FIG. 102.

strong and have a keen cutting edge.   Fig. 102 represents a good form.

Scissors for cutting filter papers, etc., will be wanted.

Hand Rolling Mills.—For assayers and jewelers, mounted upon cast iron column. The rolls are evenly tempered, truly ground and finished with a high polish.   The gears are all cut, cranks of steel, boxes of bronze, and the pressure screws of steel, with the points tempered.   They are made in at least eight sizes for hand use (and still larger ones for power), with rolls from 2x1½ to 4x2¾ inches, and weighing from 60 to 200 lbs. They cost from $30.00 to $100.00, (£6 12s. 6d. to £20 11s.)

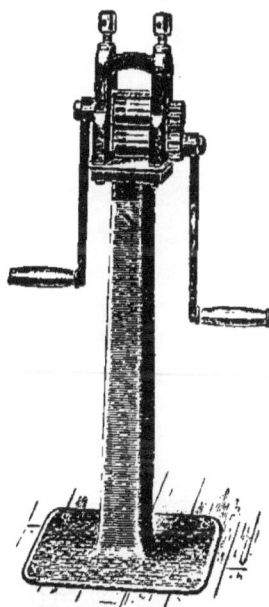

FIG. 103.

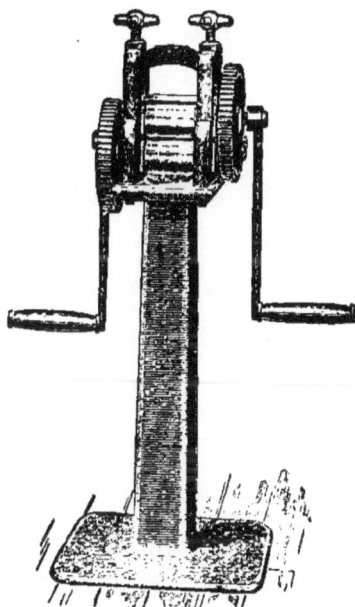

FIG. 104.

FIG. 105.

Magnifying Glass.--Pocket size very useful.

Magnet.—A small pocket magnet will come in play very often, both in the field and laboratory. Metallic iron, magnetic oxide of iron, nickel, and cobalt are attracted by it.

Figs. 106 and 107 show the two forms commonly used.

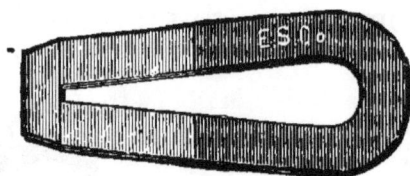

FIG. 106.

FIG. 107.

Ingot Moulds for gold, silver and lead can be obtained in many sizes and shapes. Some of the latter are herein figured.

FIG. 108.

Thus fig. 108 casts a bar $1\frac{1}{2}$x8 in., and has a slide, permitting varying lengths of bar. Fig. 109 gives an ingot $4 \times \frac{9}{16} \times 3\frac{3}{8}$, or one $7 \times 1\frac{1}{2} \times 1$, or a third $8\frac{1}{2} \times 3 \times 1$ in. Fig. 110 casts a bar, $5\frac{1}{2} \times 2\frac{1}{2} \times 2$ in., with rounded corners. Fig.

111 gives an upright ingot, $4\frac{1}{2} \times 2\frac{3}{4} \times \frac{1}{4}$ in.
Fig. 112 is for quadruple sample bars.

FIG. 109.

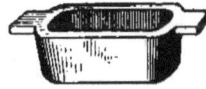

FIG. 110.

FIG. 111.

FIG 112.

Steel Alphabets and Figures.—The bullion
assayer will need these for stamping bullion.
They should comprise the numerals from 0 to
9, an alphabet, and certain
stamps in one piece, as
"Gold," "Silver," "Fine,"
"Value," "Total," "No.,"
"Oz.," and "$." In size
the above may vary from $\frac{1}{32}$
inch to $5\frac{1}{8}$ inch. Steel dies

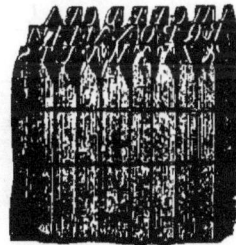

FIG. 113.

with name of mine, company, assayer, etc.,
can be procured as desired.

FIG. 114.

Bullion Punch.—Fig. 114 shows a very good punch for taking a sample from lead bullion.

Cold Chisels.—One large (1 inch diameter) and one small one (½ inch) are useful.

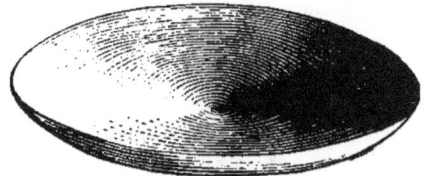

FIG. 115.   FIG. 116.

Miner's Gold Washing Pans.—Fig. 115 shows one form, material Russia or agate iron, seamless, 16½ inches diameter.   Fig. 116

FIG. 117.

gives full and sectional views of the "Batea," of wood or Russia sheet iron, and from 10 to 17 inches diameter. Fig. 117 illustrates the

"Miner's Horn," of black hard rubber, and Fig. 18 a similar one, of horn. See "Pan Test for Gold," in appendix.

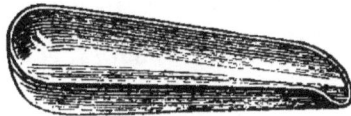

FIG. 118.

FIG. 119.

FIG. 120

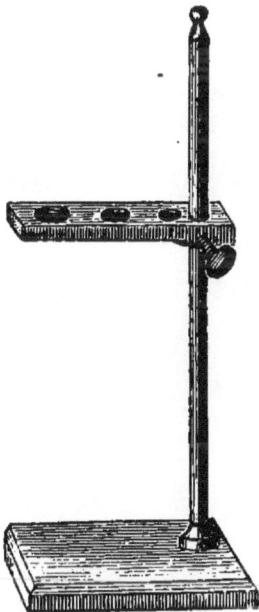

FIG. 121.

Scoops.—For ores, quicksilver or amalgam, etc. Two varieties, 5x4½ in., and 5x5½ in., both Russia iron, are illustrated. These scoops are useful implements in any assay laboratory.

Filter Stands. Fig. 121.— For holding funnels. Wooden ones are easily obtained or made. An iron ring stand can also be used.

Burette Stands. — Any simple, neat, and convenient form of support for burettes will do. A favorite one is made of iron; the clamp of brass with cork-lined jaws.

Battery, Platinum Vessels, etc.—See "Copper Analysis."

Iron Retorts.—They are used for distilling

FIG. 122.

off the mercury from an amalgam. Can be found in sizes ranging from ½ to 24 pints. The smallest size will do for ordinary work. Fig. 122 illustrates a commonly used pattern.

Chamois Skin or any other fine leather. —Used in squeezing out the free mercury from an amalgam.

CHAPTER II.

RE-AGENTS USED IN ASSAYING.

UNDER this heading I purpose to speak of those re-agents (or substances which react), necessary for the assaying of gold, silver, copper and lead ores. I shall tell what they are, how they act, when to be used and with what object, and, finally, how to prepare them when preparation is necessary.

DRY RE-AGENTS FOR ASSAYING.

The *dry* re-agents needed for the assays described in this book are seventeen in number, as follows :

1. Bi-carbonate of soda.
2. Carbonate of potash.
3. Cyanide of potash.
4. Borax glass and common borax.
5. Flour.
6. Black flux substitute.
7. Argol (or cream of tartar).

8. Common salt.
9. Carbonate of ammonia.
10. Nitre.
11. Wood charcoal.
12. Silica.
13. Lead (sheet and granulated).
14. Litharge.
15. Iron (nails and wire).
16. Silver.
17. Sulphur.

1. Bi-Carbonate of Soda (chemical name, hydro-sodic-carbonate).—This is the ordinary commercial bi-carbonate, and needs no preparation, save to be ground free from lumps. It is employed in the crucible assays of gold, silver, copper and lead ores. Its action is desulphurizing (that is, it removes the sulphur from ores fused with it, forming sulphide of soda), and oxidizing (that is, converting certain metals, as iron, tin and zinc, which may have been in the ores treated, from the metallic state to their corresponding oxides), by means of the carbonic acid it contains. Being so readily fusible, it acts as a wash to rinse down from the sides of the crucible any matters which may be adhering thereto.

Finally, it has a most important bearing as a *flux*, meaning that it forms a fusible compound with certain impurities of the ores, as metallic oxides, etc.

2. *Carbonate of Potash* (potassic carbonate). —Ordinary carbonate (*not* bi-carbonate) of potash.   Since a mixture of the alkaline carbonates (*i.e.*, carbonates of potash and soda), is somewhat more fusible than either alone, the use of this carbonate is advisable in crucible assays, particularly of gold and silver ores.   It should be ground to a fine powder and kept from the air, as otherwise it would rapidly absorb moisture.   Its action is the same as that of the bi-carbonate of soda.

3. *Cyanide of Potash* (potassic cyanide).— The cyanide which is sold in cakes can be used, after being pulverized, or, what is better, the so-called granulated cyanide, which is fine enough for all purposes.   In case that in form of cakes is on hand, it must be finely

pulverized, which ought to be done in the open air, using an iron mortar, the top of which is tied over and around with a towel. Also it is well to breathe through a wet cloth wrapped around the head across the nostrils, for the cyanide is so poisonous that inhaling the fine dust even is a dangerous practice. Use the box sieve for sifting. Keep from the air, as this salt absorbs moisture therefrom.

Employed in the lead assay. Action desulphurizing and reducing (that is, taking away the oxygen from metallic oxides and so reducing them to the condition of metals; it is the reverse of oxidizing).

4. *Borax Glass* (sodic bi-borate). — The most valuable flux the assayer possesses. He employs it both for the crucible and scorification processes for gold and silver ores, and the crucible process for copper and lead ores. It has a neutral action. The unfused borax, in powder, is often used in the lead assay.

The ordinary borax of the shops contains from 30 to 47 per cent of water of crystallization, which must be gotten rid of before the borax is fit for use. Borax, on being strongly heated, swells very considerably while losing this water, and then gradually sinks down into a clear liquid, which, on cooling, becomes the glass.

Take a large size sand crucible ("S" of Battersea make) and carefully coat its interior with either dry chalk or chalk wash. Place this in a hot fire, and drop in small pieces of borax, letting the swelling subside somewhat after each successive addition. It is well not to allow the crucible to become more than one-third full of the melted borax, as, in spite of the chalk lining, it is liable to attack the crucible and run through.

When thoroughly fused, appearing like water, pour into a frying pan coated with chalk or ruddle, and let cool. Powder in an iron mortar and sift through a 40-mesh sieve. That which goes through had best be reserved for crucible mixtures; the moderately coarse remaining on the sieve will do for scorifications.

A strong iron coffee-mill with teeth or jaws close together, will crush borax glass very finely, and in much less time than it can be done with mortar and pestle.

An iron crucible can be employed in place of the sand one. It will color the borax somewhat, which, however, does no damage.

5. *Flour.*—Wheat flour is serviceable in the lead assay, its action being reducing. But it is more commonly employed together with bi-carbonate of soda, forming what is known as

6. *Black Flux Substitute.*—A mixture of ten parts bi-carbonate of soda and three of flour. It can be used to great advantage in the crucible assays of all of our four metals.

7. *Argol* (crude bi-tartrate of potash; when pure called cream of tartar or hydro-potassic tartrate).—This is a good reducing agent, and is much used in the lead assay, and in crucible charges for gold, silver and copper ores.

The following list of the reducing powers of various reducing agents will be found to be very useful. The values are approximate only, as no two samples of any one listed will reduce to exactly the same amount, but for all practical assaying they are sufficiently near. I have included only those substances which are procurable in almost any section of the country.

TABLE OF REDUCING POWERS OF REDUCING AGENTS.

1 part of	*Will Reduce Parts of Metallic Lead*
Ordinary wood charcoal	22 to 30
Powdered coke	24
" hard coal	25
" soft "	22
Wheat flour	15
Corn starch	11½ to 13
Laundry starch	11½ to 13
Pulverized white sugar	14½
" gum arabic	11
Crude argol	5½ to 8½
Cream of tartar	4½ to 6½

See chapter III., iv., pp. 175-6 for methods of determining reducing powers of above substances.

8. Common Salt (sodic chloride).—Ordinary table salt. Very useful in every crucible assay. It serves somewhat as a protecting cover, and as a wash, bringing down from the sides of the crucible adhering metals or fluxes. If moist, place in frying pan and heat till dry, then crush free from lumps.

9. Carbonate of Ammonia (ammonic carbonate).—Of very little importance, save to assist in the roasting of certain ores. It exerts a desulphurizing action. To be employed as a fine powder.

10. Nitre or Saltpetre (potassic nitrate).—Ordinary saltpetre of commerce. Is a basic flux and oxidizing agent, and is used in the crucible assays of gold, silver and lead ores. Pulverize finely and keep dry. Determine oxidizing power as shown on page 176.

11. Wood Charcoal (carbon, more or less impure).—Very valuable on account of its reducing and desulphurizing properties. It exercises the latter action when employed in the roasting of antimonial and arsenical gold and silver ores.   Let it be in a fine condition, keep dry, and determine reducing power in same manner as for argol or cream of tartar. (See page 175.)   One part of ordinary wood charcoal will reduce from 22 to nearly 32 parts of metallic lead from litharge, according to the purity of the charcoal.   In the scorification assay of certain ores (arsenical, antimonial, etc.), charcoal exerts a beneficial action in breaking up the crust which sometimes forms on the surface of the charge.   A few pieces of roughly pulverized charcoal introduced into the matrass in parting gold bullion, excite local action and so prevent the bumping of the nitric acid solution.

There are quite a number of substances containing *carbon* in varying proportions, which, for the sake of their reducing action, might be used as substitutes for cream of tartar and charcoal, but not one of them is so effective as either of the two mentioned, and since the latter are so easily obtained, I refrain from extending the list given on page 147.

12. Silica (silicic di-oxide).—This is a valuable *acid* flux, that is, it is to be used as a flux for ores which are *basic* in character (as calc spar, dolomite, barytes, fluor spar, etc.), also for ores containing large quantities of iron oxides and carbonates and with little or no silica. It is required for the assays of certain ores of gold and silver in both the crucible and scorification processes, as will be shown. The best form in which to use it is as *pulverized silica* (sold very cheaply), since it is then in a very fine state of division suitable for intimate mixture with ores and fluxes. It should be perfectly dry.

As substitutes, in emergencies, fine, clean, dry sand can be used, and some kinds of glass (which are silicates of soda or potash, with lime, lead, etc.). Lime glass is to be preferred, but on no account is lead glass, or any containing arsenical compounds or easily reducible metallic oxides, to be employed. Common window glass and ordinary bottle glass, broken finely, will serve; and will be found to be free from objectionable metallic ingredients.

There is no advantage gained in using these substitutes, since the pulverized silica answers admirably, only it is well to know *what* to make use of, in case supplies run out.

13. Lead.—In thin sheets, called lead-foil, this metal is occasionally necessary for cupellations, as described under the assaying of gold and silver, and in the gold bullion assay. It should be tested for silver. (See page 174).

In the granulated form (when it is sometimes called test lead) lead is as invaluable as borax glass for the scorification assay. It can be purchased of varying degrees of fineness and purity, or it can be made from bar lead by the assayer himself, as here directed.

Melt pieces of the bar lead in any convenient vessel (odd sizes of sand crucibles, for instance), and when it is of a temperature just hot enough to char a splinter of wood, pour into a compactly-joined cigar box without a cover, or a strong starch box. Immediately give a gentle rotary motion to the contents of the box, till the lead begins to thicken, and emits a slight creaking noise, when the motion is to be increased to a final vigorous shaking from side to side. A minute or two of this latter, and the thing is done. Sift through a 20-mesh or an ordinary flour sieve, and remelt that which remains on the sieve. When the entire batch has been thus granulated, assay for silver, following the directions on page 172.

14. Litharge (plumbic mon-oxide, yellow oxide of lead).—Employed mainly for the crucible assays of gold and silver ores. It should be dry, and free from any considerable amount of red oxide of lead, as this causes oxidation of silver, and consequently loss. Mitchell says: " Ordinary litharge can be easily freed from this oxide by fusing it and pouring it into a cold ingot mould, then pulverizing, and carefully keeping it from contact with air, as it readily absorbs oxygen, and if it be allowed to cool in the atmosphere, it will nearly all be converted into the red oxide."

Litharge can quite easily be procured free from large quantity of red oxide, and if it is kept in a tightly-stoppered bottle or tin can with closely fitting cover, there is little danger of conversion to this oxide.

It is used to furnish metallic lead that serves as a solvent for the precious metals in the ore. When in the melted state it has the power of giving up its oxygen to almost all the metals (save gold, silver and those of the platinum group), converting them into oxides, and since these are generally extremely fusible, they go into the slag. Thus we are able to separate gold and silver from any baser metals they may be combined or associated with.

Litharge is a very powerful desulphurizing agent (see Mitchell, pp. 181 to 187), and also serves as a metallic flux.

It may safely be stated that *all* litharge contains silver to a greater or less degree.

It may be, and generally is, in small quantity, but it is absolutely necessary to determine the amount, and to allow for it in the calculation of silver in any ore tested.

For the determination of this, see page 168.

White lead (ceruse, plumbic carbonate, or carbonate of lead), and sugar of lead (plumbic acetate), can be made use of as substitutes for litharge, but they do not act quite so well.

15. Iron.—A good desulphurizing agent, and as such is much employed in the assay of galena or sulphide of lead. Wire of $\frac{1}{16}$ inch diameter, and eight-penny nails are the correct sizes. Iron filings can sometimes be used.

16. Silver.—Can be bought as very thin foil. It is quite often needed in *inquartation* (which see). It should be tested for gold by dissolving $\frac{1}{2}$ gramme in pure nitric acid. After the solution is complete, there should be no black specks (gold), no matter how small, in the liquid. There is generally no difficulty in procuring silver perfectly free from gold.

17. Sulphur.—Ordinary commercial sulphur. It is *the* sulphurizing agent. Used in Aaron's crucible method to form mattes.

(Oxide of iron and iron pyrites may be wanted for the assay of refractory copper ores—they need no especial description.)

WET RE-AGENTS FOR ASSAYING.

1. Distilled Water.—A most satisfactory, simple and efficient device for making it is the "Domestic Water Still," made in two sizes, and costing $15.00, (£3 2s.) and $25.00, (£5 3s.) respectively. Figs. 123 and 124 give full and sectional views. To set up, screw the apparatus to a wall, over a sink. Attach a piece of rubber hose to the small faucet "B," at bottom and connect to faucet of water-pipe. The lower part of upper tube is the over-flow and should have a rubber hose connected and leading to the sink. A third hose attached to corrugated end of burner under the still, supplies the gas. The thumb-screw at bottom of still affords entrance to the still for cleaning purposes. The dis-

tilled water drops from the spout " Z." Light
and regulate water supply until the waste water
runs tepid.

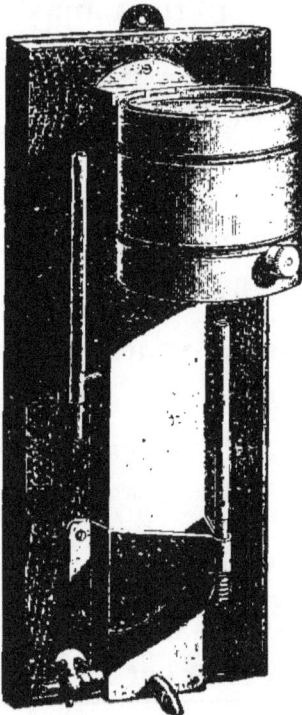 

FIG. 123.   FIG. 124.

Pure rain water is a very fair substitute.
The main point in any case, is to see that it
contains no *chlorine* (indicating usually,
chloride of sodium or common salt). Test
the water for this element by acidulating a
clear sample with *pure* nitric acid, and adding a

drop or two of nitrate of silver solution (made by dissolving one part of the dry nitrate of silver in twenty parts of distilled water). The water should remain perfectly clear, that is, there should not be in it the slightest cloudiness or turbidity. If it does show this, reject, and prepare or secure a fresh supply.

2. *Nitric Acid* (hydric nitrate).—Indispensable for *parting*, *i.e.*, the separation of silver and gold by dissolving out the former. It can be procured perfectly pure, but should always be tested for *chlorine*, in same manner as is distilled water. Should it contain this objectionable ingredient, it can be removed by adding *one drop* of nitrate of silver solution and letting the acid stand in the light till the purple-black precipitate of chloride of silver settles to the bottom of the bottle. Then add a second drop, and let remain undisturbed as before. Continue these successive single drop additions until finally the last drop ceases to form any precipitate or milkiness in the acid. Draw off the clear acid and keep tightly stop-

pered. There are two reasons why chlorine should not be found in the nitric acid. First, it will tend to throw down, as silver chloride, the silver dissolved out of a bead by the nitric acid in the process of parting. Secondly, it indicates the presence of hydrochloric acid, and this acid forms *aqua regia* with the nitric acid, which could easily dissolve the very small amounts of gold sometimes left after parting.

RE-AGENTS FOR ANALYSIS.

The other re-agents, wet and dry, used in the qualitative tests, analyses, and special processes, are the following :

Acetic Acid.— Needed in qualitative tests. Should be pure. Dilute with two parts distilled water.

Alcohol.— Wanted in the copper analysis and perhaps as fuel for a lamp. Use either common alcohol or wood spirits.

Ammonia Water (ammonic hydrate, caustic ammonia, aqua ammonia).— If very strong, dilute one part with two parts of distilled water.

Bi-chromate of Potash (potassic di-chromate).—Used in the volumetric determination of iron. Should be procured pure.

Black Oxide of Manganese (manganese dioxide).—Necessary to aid in the preparation of chlorine gas. Does not need to be perfectly pure.

Bromine.—Used to remove manganese from its solution, by precipitation, in the volumetric copper analysis. Agitate some of the liquid bromine with distilled water in a glass-stoppered bottle, and use the resulting aqueous solution for the precipitation.

Carbonate of Ammonium (ammonic carbonate, carbonate of ammonia).—Needed in the volumetric copper analysis. Should be procured chemically pure.

Carbonate of Sodium (sodic carbonate, carbonate of soda).—Used in the volumetric copper and other analyses, to precipitate iron, manganese, blow-pipe test for manganese, etc.

Should be pure. Dissolve in ten parts distilled water.

Caustic Potash (potassic hydrate).— One part of common stick potash, dissolved in ten parts of water.

Caustic Soda (sodic hydrate). — In stick form. Dissolve in distilled water — four parts — when wanted.

Chloride of Barium (baric chloride, muriate of baryta).— One part of the pure salt dissolved in ten parts of distilled water.

Chloride of Calcium (calcic chloride).—The dry, fused lumps, used to keep moisture away from fine scales. Need not be chemically pure.

Citrate of Ammonium (ammonic citrate).— Dissolve one part of the salt in ten parts of distilled water.

Citric Acid.— Pure, for volumetric copper and other analyses — to keep iron in solution.

Cyanide of Potassium (potassic cyanide, cyanide or prusside of potash).—Pure, for the volumetric copper analysis.

Ferrocyanide of Potassium (potassic ferrocyanide, yellow prussiate of potash).—One part of the pure salt, dissolved in twelve parts of distilled water.

Hydrochloric Acid (muriatic acid).—To be pure. One bottle may be of the concentrated, a second of a mixture of one part acid with four parts of distilled water.

Hyposulphite of Sodium (sodic hyposulphite or thiosulphate).—The pure salt is employed for the volumetric determination of manganese, the chlorination test for silver, and as a precipitant for copper in the volumetric analyses of the ores of this metal.

Iodide of Potassium (potassic iodide, iodide of potash).—Wanted in the volumetric determination of manganese and as a test re-agent for lead. When used for the latter purpose, it may be either in the solid form, or in solution in water — one part in ten.

Lime Water (calcic hydroxide).— Place a very little slaked lime in a bottle; fill with water and shake. Keep tightly corked, and, when wanted, draw off the clear liquid without disturbing the sediment.

Mercuric Chloride (corrosive sublimate).— Needed only for the volumetric determination of iron, which see.

Metallic Copper.— Wire for battery purposes, sheet for amalgamation test in panning, and some *pure* to form test solutions in the volumetric copper analysis, will be needed.

Metallic Iron.—Pure, to precipitate copper from its solution, in the volumetric analysis of the latter metal.

Metallic Mercury (quicksilver).—Some that is impure can be employed to amalgamate the zinc plates of a battery, and some *free from gold and silver* will be wanted in the various amalgamation tests.

Metallic Zinc.—In plates, forming a part of a battery. As a re-agent, zinc in pencils, or granulated, will be needed *pure*.

Nitrate of Silver (argentic nitrate, lunar caustic).— See testing of distilled water for chlorine.

Nitric Acid.—A bottle of pure and concentrated acid, and one of the common commercial (concentrated) for battery, should be on hand.

Stannous Chloride (proto-chloride of tin, "tin salts").—For the volumetric determination of iron.

Sulphate of Iron (ferrous sulphate, green vitriol, copperas).—In solution in water (of no particular strength) it is used to precipitate gold from its solution as a chloride, after the chlorination assay.

Sulphate of Magnesium (magnesic sulphate, sulphate of magnesia, "Epsom salts").—In a pure state, to precipitate arseniates in the volumetric copper analysis, as a test re-agent for phosphates, etc.

Sulphide of Iron (ferrous sulphide, sulphuret of iron).—See next paragraph but one. Can be purchased, or made by holding roll sulphur against a bar of red-hot iron.

Sulphocyanide of Potassium (potassic sulphocyanide).—One part of the pure salt dissolved in ten parts of distilled water.

Sulphuretted Hydrogen Water (hydrogen

sulphide gas dissolved in water).—Very useful in qualitative analysis. To generate it, fit
together a simple piece of apparatus similar
to fig. 125. The larger bottle, which may be
of any capacity above six ounces, is provided
with a doubly-perforated cork, through one
hole of which passes a straight glass tube to
nearly the bottom of the bottle, and terminating in a funnel. Through the other hole a
second tube passes a little way into the larger
bottle, and bending twice at right angles, goes
through the cork of the smaller bottle to
nearly its bottom. A third tube leaves this
smaller bottle and connects by a bit of rubber tubing with a fourth tube dipping into
the receiving bottle containing distilled water.
Place an ounce or two of sulphide of iron
broken in small pieces in the bottom of the
large bottle and fill half way up with ordinary water. The small bottle is to be half-
filled with distilled water to wash the gas.
Pour some common sulphuric acid into the
funnel-tube, when the gas will at once be

given off. To ascertain when the water in the re-agent bottle is saturated, hold the thumb tightly over its mouth, and shake. On releasing the pressure a little the thumb will be held down if the water is not saturated, but will be forced up, if the contrary is true.

A little glycerine put in the re-agent bottle will help to retain the gas in solution.

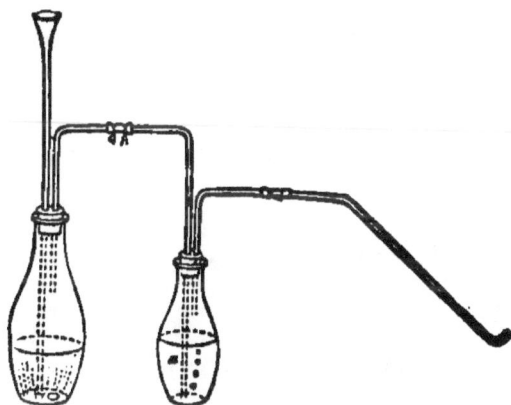

Fig. 125.

Sulphuric Acid (oil of vitriol).—A bottle of pure and another of common, both concentrated. If dilute acid is wanted, mix, in a beaker, one part of the acid with five of distilled water.

Tartaric Acid.— Used to retain iron,

alumina, etc., in solution, in the volumetric copper analysis and in qualitative analyses. The pure salt can easily be procured.

MISCELLANEOUS.

Bone-ash.—For making cupels, which see. It is best to use a good quality.

Chalk and Chalk Wash.—Ordinary chalk, to be used dry, and the same finely ground and rubbed up with water, for coating crucibles, etc.

Clay Lute.—Fire-clay and sand, with solution of common borax in water to bind them together. Horse and cow-hair may also be mixed with them.

Ruddle (ferric sesqui-oxide, red oxide of iron, hematite).—A lump for marking cupels and scorifiers, and a paint (prepared by putting an ounce or two of the fine powder with water in a bottle, and shaking) for marking crucibles, coating frying-pan, etc., are wanted.

CHAPTER III.

TESTING OF RE-AGENTS.

BEFORE proceeding to make the regular assays, the student will find it expedient to examine his re-agents, either to ascertain the presence or absence of silver (and in the former case, to determine its quantity), or to learn their various strengths, as shown in their reducing or oxidizing powers.

The following five divisions include all the requisite tests of re-agents :

    I. Testing of Litharge for Silver.

   II. Testing of Granulated Lead for Silver.

 III. Testing of Sheet Lead for Silver.

 IV. Determination of the Reducing Powers of Reducing Agents.

  V. Determination of the Oxidizing Power of Nitre (Nitrate of Potash).

I. TESTING OF LITHARGE FOR SILVER.

As stated in the chapter on re-agents, almost all litharge contains silver, generally as a small amount. However minute this may be, we must know exactly what it is, and allow for it in calculating the value of an ore.

This we do by the crucible process, in the same manner as we should run an ore. (See Part II, Chapter I.)

Mix *very thoroughly* the particular lot of litharge to be examined, and sample as usual.

Make the charge

	Assay ton weights.	*Gramme weights.*	*Grain weights.*
Bi-carb. soda..........	½ A. T 15	grammes..	240 grains = ½ oz.
Carb. potash...	¼ " 7½	" ..	120 " = ¼ "
Litharge...	1½ " 45	" ..	720 " = 1½ "
Charcoal, ⎰ Any ⎱	½ gramme ...	½ gramme.	7½ "
Flour, ⎰ one of ⎱	1 "	1 " ..	15 "
Argol, ⎱ these. ⎰	2 grammes...	2 grammes..	30 "
Salt cover.			

Any one of the above charges will produce a lead button of from 15 to 20 grammes (131 to 308 grains).

Mix everything well, and brush into an " S "

* See Chapter I, pp. 66-70, for full explanation of these weights.

Battersea crucible or its equivalent (4¾ inches by 4⅛ wide).

Have the fire quite hot, and heat crucible till contents are in quiet fusion, which will be in from twenty-five to thirty-five minutes. Take out, let cool, break, and hammer button into shape.

If the button is too large for any cupel, reduce by scorifying, then cupel. (See " Scorification and Cupellation," Part II, Chapter I.)

Weigh the resulting silver button, and deduct its weight from the gold and silver bead obtained from any crucible assay *where the same quantity of litharge has been employed.* If more or less than 1½ A. T. (or its equivalent in grammes or grains) is used, calculate and deduct accordingly.

For example, one lot of litharge I have tested carried 0.75 (¾) of a milligramme for the 1½ A. T., which amount was made the factor for that particular lot.

The above amount, equivalent to ½ ounce per ton of 2,000 pounds, is of course very small, and, in the calculation of the value of an ore running say 100 oz. and upward, need not be deducted from the weight of the silver bead, since the loss of silver by absorption and volatilization from such a bead while cupelling, would more than counterbalance it. But it is very important that it should be deducted in the case of a poor ore, and especially when there is a question as to the presence or absence of silver in any ore.

For the sake of practice, it will be well for the student to perform this crucible assay of litharge three or four times.

Many assayers, and particularly those newly entered into the profession, continually try to obtain litharge (and for that matter, granulated lead) free from silver, meaning that it shall contain absolutely no silver. But, in the first place, they cannot procure it, and, in the second place, if they could, there would be nothing gained, in my opinion, by using such, as I will endeavor to show.

First, in assaying ores which might be rich

in either gold or silver, it would make no difference whether the litharge employed contained absolutely no silver, or the small
amount it usually carries, for in the latter case
its weight would not be deducted from that
of the button for reasons just given. Secondly, in ores very low in gold, it becomes
very difficult, and sometimes impossible, to
find the minute speck of gold left from the
assay (particularly as the result of a scorification assay), even with the aid of the magnifying glass ; but when the litharge does contain
a little silver, the latter not only leaves itself
and the gold together in a visible and tangible
form on the cupel, but it also serves to collect
the gold during the process of crucible fusion,
and retains it always thereafter. Thirdly, a
small but known amount of silver in litharge,
tests the assayer, his methods and practice,
the litharge itself and some of the ores worked
upon, for he ought to get the constant figure
of the silver in the litharge when testing the

many worthless ores he is bound to examine in the course of his work.

II. TESTING OF GRANULATED LEAD FOR SILVER.

As in the case of litharge, all granulated lead must have its amount of silver determined, which is done by the scorification process.

Mix and sample as usual.   Rub the interior of the scorifier with a little fine silica before pouring into it any of the lead ; it will serve to protect the scorifier from corrosion by the molten lead.   Weigh very carefully 2 A. T., 60 grammes or 960 grains (2 oz.), of the lead, and pour into a 2¾ inch scorifier, and deposit on the top a piece of borax glass about the size of the head of a pin.

Scorify and cupel as shown in the next chapter, and weigh resulting bead.   The weight of the silver bead, divided by two, will give the number of milligrammes or fractions, that one assay ton (30 grammes, 480 grains, or 1 ounce) of the lead contains, which I have found to vary from $\frac{3}{10}$ milligramme to 1.2 milligrammes.

Make a table of the amounts of silver contained in fractions and multiples of one assay ton, and post it in some convenient place for reference. I give an example of one particular lot :

0.50 A. T. contains 0.40 milligramme silver.
1.00 " " 0.80 " "
1.50 " " 1.20 " "
2.00 " " 1.60 " "

If other weights of lead are used, calculate accordingly.

Deduct silver, in proportion due to the amount of lead used, from beads coming from ores ranging less than 100 ounces; above that disregard it, as with litharge. As with litharge, make several runnings of the lot of lead.

The reasons given for desiring a litharge with some little amount of silver present are almost equally applicable to granulated lead.

III. TESTING OF SHEET LEAD FOR SILVER.

Sheet lead can generally be purchased remarkably free from silver, and as it is seldom that a piece of more than ten or twelve grammes (120 to 150 grains) in weight is required, the quantity of silver such a piece will contain will be exceedingly small. Moreover, its chief use being to enwrap gold and silver beads for recupellation, this amount of added silver is too minute to counterbalance the loss of silver by volatilization and absorption. Sheet lead is sometimes employed to aid in cupelling gold beads that have been inquarted, and here a loss or addition of silver is not important.

But should the lead-foil be suspected of carrying any quantity of silver, its exact amount can be determined by cutting off from various parts of the foil, and in small shreds, two or four assay tons (60 or 120 grammes, 2 or 4 ounces), which are to be scorified and cupelled as usual.

IV. DETERMINATION OF THE REDUCING POWERS
OF REDUCING AGENTS.

1. Argol (p. 146).—Weigh out the following charge:

Bi-carbonate of soda..15 grammes, 240 grains (½ oz.)
Carbonate of potash.. 7½   "   120   "   (¼ ")
Litharge............45   "   720   "   (1½ ")
Argol.............. 2   "   30   "
Salt cover.

Put into a small crucible (size " V " of Battersea), place in a hot fire, cover, remove when thoroughly fused, cool, detach button from slag, weigh, following the directions given for the crucible assays of gold and silver.

The result, divided by two or thirty, will give the number of parts of metallic lead that one part of argol is able to reduce from litharge. It ranges around 8.5 parts.

2. Cream of Tartar, or bi-tartrate of potash (p. 146). — Same charge as above, excepting that the two grammes or thirty grains of argol are to be replaced by three grammes or forty-five grains of the tartar. One part of pure tartaric acid will reduce 6 parts of lead, and the same amount of ordinary cream of tartar will reduce 6.4 parts of lead.

3. Charcoal (p. 148).—Make up this charge :

Bi-carb. of soda..2 A. T., 60 grms., 960 grains or 2 oz.
Carb. of potash ..¼ " 7½ " 120 " " ¼ "
Litharge ....2 " 60 " 960 " " 2 "
Charcoal ....1 gramme or 15 grains.
Salt cover.

Use an " S " crucible. As previously stated, the reducing power of one part of charcoal varies between 22 and 32 parts of lead.

The reducing powers of the other substances given on page 147 are determined in a similar manner to the three quoted above, running coals and coke as for charcoal, white sugar as for cream of tartar, flours, starches, etc., as for argol, etc.

V. DETERMINATION OF THE OXIDIZING POWER OF NITRE (NITRATE OF POTASH).

Determine the oxidizing power of the fine, dry salt (p. 148) by the following charge :

Bi-carb. of soda..2 A. T., 60 grms., 960 grains or 2 oz.
Carb. of potash ..$\frac{1}{4}$ " $7\frac{1}{2}$ "   120   "   " $\frac{1}{4}$ "
Litharge ....2 "   60 "   960   "   " 2 "
Charcoal ....1 gramme or 15 grains.
Nitre......... ...5 grammes or 75 grains.
Salt cover.

Use an "S" crucible, and treat as in the previous crucible operations. The difference between the weight of the lead button obtained and that found in the assay of the charcoal, divided by five or seventy-five, will give the oxidizing power of the nitre, per part. It is about four parts.

PART II.

ASSAYING.

PART II.

ASSAYING.

CHAPTER I.

GOLD AND SILVER ORES.

OCCURRENCE.—Gold is found in large quantities in the native state, designated by the various names of free gold, flour, leaf, wire and nugget gold. The minerals which most frequently carry gold are oxide of iron, pyrites of iron and copper (known as auriferous sulphurets), arsenopyrite, and tellurium ores; of these, the most abundant are the first two.

Minerals which less frequently are gold-bearing, are galena, blende, gray copper and "carbonate ores."

For a classification of silver ores I quote from Kustel's "Roasting of Gold and Silver Ores":

"IMPORTANT SILVER ORES.

The most important silver ores are those found in such quantities as to be an object of metallurgical operations. The principal minerals of this kind are the following :

A. Silver ores with unvariable amount of silver. — a. Sulphuret of silver, or silver glance, with 87 per cent of silver. It is of common occurrence. *b. Brittle silver ore* (*stephanite*), or sulphuret of silver and antimony. This mineral contains 68 per cent of silver, and is quite common. *c. Polybasite*, sulphuret of silver, antimony and some arsenic, with 75 per cent of silver. *d. Ruby silver*. The dark red silver ore, or antimonial variety, with 59 per cent, and the light red silver ore, or arsenical variety, with 65 per cent of silver, are valuable minerals. *e. Miargyrite*, sulphuret of silver and antimony; 36.5 per cent of silver. *f. Horn silver*, or chloride of silver, with 75 per cent of silver. *g. Iodic* and *bromic* silver of yellow and green colors.

B. Argentiferous ores with variable amount of silver.—*a. Stromeycrite,* or silver copper glance, a sulphuret of silver and copper containing up to 53 per cent of silver. *b. Stetefeldite,* with 25 per cent of silver, is an oxide ore. *c. Silverfahlore,* argentiferous gray copper ore. It contains silver in very variable proportions up to 31 per cent. This ore is quite common, and for this reason is important. It is also one of the most rebellious ores, containing copper, antimony, arsenic, sulphur, lead, iron, zinc, and sometimes gold and quicksilver. *d. Chloride ores* (so-called), mostly decomposed ores, generally of an earthy appearance and different colors. They contain more or less finely divided chloride of silver.

C.—*a. Argentiferous lead ores,* galena, or sulphuret of lead, lead glance. Generally, this is not rich in silver, containing from $20 to

$60 per ton.   Specimens assay sometimes as high as $300.* *b. Cerussite*, carbonate of lead.   If pure, without admixture of copper and other carbonates, it is poor in silver in most cases.   *c. Argentiferous zinc blende*, sulphuret of zinc.   Pure zinc blende contains usually only traces of silver ; often, however, it assays well, even up to $400 per ton.   *d. Argentiferous pyrites.*   Copper and iron pyrites are poor in silver, but often auriferous.

There are, besides, numerous classes of decomposed silver ores, generally of earthy nature ; also, half decomposed ores which have lost their metallic glance, having a black or bluish-black color, and being generally cupriferous." †

ASSAY.—We can best consider the systematic fire treatment of gold and silver ores, by dividing it into a series of operations, and taking each in turn and in detail.

* I have found them as high as $1,500 per ton.   (W. L. B.)

† See appendix for extended lists of the minerals of or containing gold and silver.

The three main divisions are :

I. Preparation of the sample.

II. Scorification process.

III. Crucible process.

I. PREPARATION OF THE SAMPLE.

The first thing to be done in the treatment of an ore, whether it is to be assayed for gold, silver, copper, lead, or any other metal, is to *place* it, that is, to label it.   This is best accomplished by giving to it a running number, never to be repeated.   By adopting this system of numbering all samples, any danger of confusing specimens from various mines or parts of the same mine or vein, is entirely gotten rid of.   Have a notebook at hand, and, in it, under the number, write such items as may be necessary or useful, as the date when sample was received, name of person sending it, character of the ore, nature of the charge, weights employed, calculations, etc. To pieces of the ore which are to remain whole, affix gummed labels, bearing the same number. To preserve the final pulverized samples, bottles of about four ounces capacity, cork-stoppered, and similarly labelled, can be employed, or what is even better, pasteboard boxes in size about $5\frac{1}{2}$ inches long by $3\frac{1}{2}$ inches wide and 2 inches high, will be found to be very serviceable. They can be written on, thus requiring no labels.

The next step is to secure an *average sample* for assay, and its importance cannot be over-rated. An ore is by no means of uniform character, being, in general, made up of the gangue or valueless portion of the ore, through which are scattered the valuable minerals. Therefore, unless the sample finally chosen for assay represents an average of the entire lot, being a mixture in the same proportions, of the richest, the medium and the poorest portions, as in the original ore, the assay itself is worthless, no matter how carefully it may have been performed.*

* In this connection, I would refer the student to an article "On the Commercial Sampling of Minerals," by Mr. L. S. Austin, of Salt Lake City, Utah, which appeared in the "Engineering and Mining Journal" (July 22, Aug. 5, Aug. 26, and Sept. 16, 1882).

To illustrate the averaging, take a quantity of ore weighing fifty pounds, which may be as a single lump, or, better, the result of the selection of samples across a section of the vein. In order to get a fair average, it is not necessary to operate on a larger quantity than this amount, for above it, should come in, as a more practical test, the mill-run.

With a heavy sledge-hammer, break up the entire mass into pieces of about the size of a hickory-nut. Should the rocks be so large or so very hard as to obdurately resist the hammer, they may be brought into submission by that process called "astonishing," by Prof. Chapman, in his valuable little work. It consists in heating red-hot the resisting pieces, and plunging them while thus into some cold water in a pan. This heroic treatment will either at once reduce the lump to small pieces, or render it so friable that light pounding will pulverize it. Pour off the water and dry the contents thoroughly, then break into small pieces as directed. Transfer to a large sheet of heavy brown or manilla paper, then, with a large iron or steel spatula, thoroughly mix, by turning over and over and by stirring in together with the dust, the finer and coarser particles, till satisfied that the whole is a homogeneous mixture.

(At this stage of the operation it is a good plan to reserve a characteristic lump or a few pieces, from an examination of which the nature of the ore may be determined, and process of treatment decided upon.)

Now divide in halves by means of a very large spatula or piece of heavy sheet brass, 18 inches long by 4 inches wide, and $\frac{1}{8}$-inch thick. Break up still finer (to the size of a hazel-nut or less) the half selected. Mix again and halve as before. Continue the crushing, mixing and halving until about one pound has finally been sampled down.

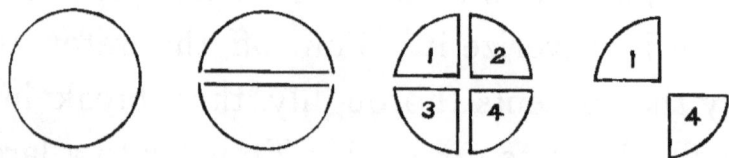

FIG. 126.

Instead of halving, the piles may be quartered, and two of the diagonally opposite quarters taken as a half. Fig. 126 shows the pile entire, divided, quartered, and two of the quarters removed; parts 1 and 4, or 2 and 3, are to be put together.

When the ore to be assayed is less than fifty pounds, ranging down to a pound or two, it can be broken still smaller in the successive steps, and when it is but a few ounces in weight, the whole of it should be crushed and pulverized, as directed. Wet or damp ores and pulps should be dried before pulverizing.

The student must exercise his judgment in a measure, with regard to the sampling of an ore, simply remembering that the object, as before stated, is to obtain a final product which shall be an exact counterpart, in relative proportions, of the metals and gangue of the original ore.

Instead of halving, the broken ore may be taken up on a sampling shovel, and thrown on a tin or copper sampler, making it a rule to reject either all that which goes between the prongs or ribs, or that which remains upon them.   These two articles are convenient, but not necessary.

The third step is to pulverize the sample finally obtained, which may be done very simply though somewhat laboriously (depending considerably upon the nature of the ore), by means of an iron mortar and pestle.   A towel wrapped loosely around the pestle and across the top of the mortar will prevent loss due to flying particles.

Sift through a sieve of eighty or ninety or even of one hundred meshes, since the finer the powder, the more quickly will it be acted upon in the furnace.   Ores which contain much clay or lime are very apt to clog the meshes of the sieve so that little will pass through.   This may generally be obviated by placing in the sieve an ounce weight or equally heavy piece of smooth iron, the movement of which in shaking keeps the meshes open. Do the sifting over a piece of brown paper, and be sure that *all* the sample passes through the sieve, for the few minute particles or scales, that might remain on the sieve and be hastily thrown away, could be of sufficient value to vitiate the assay. I might give here, however, a little suggestion made me by Mr. H. H. Corbin, of Telluride, Colo. Certain ores contain free silver (with a little gold in it, sometimes), to such an extent that while they will not produce many scales, yet there may be a few left on the sieve. In such cases dump the scales on the grinding plate and cover them with the finely-powdered ore which has already gone through the sieve. Grind heavily for quite a little while, then sieve again. It will be found that most of the scales have been ground by the ore (particularly when the latter is quartzose), fine enough to go through a 100-mesh sieve. If they will not, repeat the treatment until they will. Three grindings will usually suffice. The sievings are to be *very carefully* mixed. Even all this trouble is less than that of the ordinary scale method of treating ores containing the precious metals in the free state. (See in appendix, " Assaying of Ores containing Free Gold or Free Silver".)

Mix again the fine powder, and with a large brush transfer to the properly marked box or bottle, when the sample is ready for assay. Do not shake the box or bottle between the times of grinding and weighing, as this tends to cause the gold, silver, sulphurets, or other heavy minerals to settle, resulting in an unequal distribution of the various constituents of the ore. If much time elapses between the weighings and the second treatment the ore should be remixed.

When very many assays have to be performed daily, the rubbing-plate and rubbers will be found so very convenient and so time-and-labor-saving, that they will become almost necessities. (See pp. 29–32 for description.)

The operation of grinding, or rubbing, or pulverizing, is managed as follows : The ore, previously broken into fine pieces as directed, is placed upon the *clean* surface of the plate, the rocker is now laid upon it, and with one hand firmly pressed upon the body of the rocker, and the other grasping its handle, it is moved backward and forward with an oscillating motion. This knack of grinding, although not easy to describe, is soon acquired.

After grinding, the plate should be thoroughly cleaned, which can be done by rubbing on it either sand, quartz, broken glass, common salt, old cupels or scorifiers, broken crucibles, the slags from scorifications, or some worthless ore, finishing with an old rag. Very great pains should be taken to perfectly clean the plate after grinding much free gold ore or such rich ores as tellurides.

In place of the sheets of brown paper already mentioned, and which quickly become full of holes, the zinc sifting-pans (p. 41, fig. 18) or a rubber cloth, can be used to advantage.

To guard against loss of dust, the tin box-sieve (p. 40) is recommended. There is another laboratory plan for systematic sampling, which is very fair, though slow and best suited for small lots of from 5 to 20 pounds of ore. For this the assayer needs four sieves in addition to the fine one—they are 2, 4, 8 and 16-mesh, respectively.

Crush the sample so that it will just about go through the 2-mesh sieve, mix thoroughly, divide in halves, reject one-half, and crush the remainder so it will just pass through the 4-mesh sieve. Again mix, divide in halves, reject half, and crush the other half for the 8-mesh sieve. Mix, halve, reject half, and crush for the 16-mesh sieve. The half kept from this is all to be pulverized and put through the 100-mesh. All this insures thorough mixing and uniform crushing and sampling.

Having now finely ground the ore to be assayed, we must next decide how to treat it.

There are two methods of assaying gold and silver ores, the scorification and the crucible.

The process to be chosen depends chiefly

on the nature of the ore. In general, we may say the scorification process is better adapted for *all* silver ores, and for *rich* gold ores (including telluride ores of any degree of richness).

The crucible process serves better for *low grade gold ores.* The advantage of this process lies mainly in the fact that it enables us to operate upon a larger quantity of ore; otherwise it is no better than the scorification method and indeed in many respects the latter is to be preferred.

The scorification process is so much simpler to use, easier to comprehend, and so satisfactory in its working, that I shall give it the first place in this manual.

II. SCORIFICATION PROCESS.

The object of this process is to so act upon an ore with heat, access of air, and certain re-agents, that the precious metals shall be driven out of their combinations with the impurities of the ore (or if free, separated from them), and be retained alloyed with another metal, lead, and from which they can afterwards be separated.

The chief re-agents are lead, in a granulated condition, and borax glass.

Besides these, silica, iron, and bi-carbonate of soda are occasionally employed.

The ore, mixed with the lead, and covered with the borax glass or other flux, is put into a scorifier and subjected to heat in a muffle.

Under the action of the heat, the lead melts, and being scattered throughout the ore, seizes upon the gold and silver and settles with them to the bottom of the scorifier. The borax glass or other flux attacks the gangue and impurities present, and uniting with them and with litharge resulting from oxidation of some of the lead, forms a slag or glass, which floats upon the surface of the molten lead.

So much for the theory of scorification; in practice we follow in regular rotation the steps here given:

a. Preparation of Charge (including weighing of ore, roasting, weighing of re-agents, mixing, etc.)

b. Scorification.

c. Cupellation.

d. Weighing the Gold and Silver Bead.

e. Parting.

f. Inquartation.

g. Weighing the Gold Residue.

h. Calculations.

a. Preparation of Charge.

Whatever subsequent treatment (including roasting) an ore is to undergo, the amount required for assay must *always* be weighed first. If the ore is in a box, it can be sampled therein, but it is better to pour from it or from the sample bottle onto a clean piece of black glazed paper or sheet rubber, and with a spatula form it into a smooth square. Divide into smaller squares with a rule and take a dip from each division, as shown in sketch.

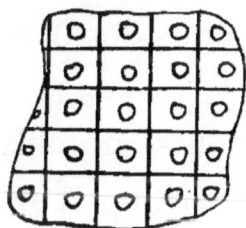

Fig. 127.

Have ready cleaned a number of the scorifiers. Number or letter each scorifier with ruddle (liquid or lump), weigh the requisite amount of granulated lead, divide approximately in halves, and transfer one-half to the scorifier. Upon it brush the ore (roasted or not) and mix by means of a small steel spatula. Pour the remaining half of the lead evenly over the surface of the mixed ore and lead, and over all sprinkle the borax glass. In similar manner prepare all the other charges.

A deviation from this method has been followed by some assayers. Their procedure is to put, say, $\frac{1}{4}$ of the lead at the bottom of the scorifier, then the ore and $\frac{1}{2}$ of the lead mixed previously, topping all with the remaining $\frac{1}{4}$ of the lead. This change is due to the fear of unacted-upon ore remaining at the bottom of the scorifier. I am inclined to consider it an almost unnecessary refinement.

In many assaying establishments, notably the larger ones, the practice of marking scorifiers, cupels, and crucibles does not obtain.

Instead of this, a systematic order of arrang-
ing these articles is kept up, either in or out of
the furnace, and this routine of position and
order of working is never varied, so that by
relative place a sample can always be identi-
fied. This plan is indeed a good one, and
perhaps imperative where very much work is
done, but for the beginner, for a time at least,
the custom of marking everything had better
be adopted.

CHARGES.

(For manner of roasting see " III. Crucible Process.")

1. One for " every day " ores, serving well
for the common run of ores in which the
metals are not in excess of the gauge, is the
following:

Ore. .... $\frac{1}{5}$ A. T., 5 grammes, 96 grains or $\frac{1}{5}$ oz.
Granulated lead ...... 1$\frac{1}{2}$ " 45 " 720 " 1$\frac{1}{2}$ "
Borax glass.........250 mgrms., or 4 grains.

Use it for ores that do not contain much
copper or lead.

2. Copper glance or copper pyrites :

Ore .... $\frac{1}{10}$ A. T., 2$\frac{1}{2}$ grammes, 48 grains or $\frac{1}{10}$ oz.
Granulated lead ... 2$\frac{1}{2}$ " 75 " 1,200 " 2$\frac{1}{2}$ "
Borax glass .. .. 200 mgrms., or 3 grains.

For the above ores, a preliminary roasting can be made, if considered advisable.

If not to be roasted, and when there is not very much of the sulphurets present, heat gently for a time, till the roasting in the scorifier is done.

If rich in sulphurets, a strong heat can be applied at once, melting everything down into a sort of matte, then proceeding as usual.

3. Copper matte:

Matte ....$\frac{1}{10}$ A. T., 2$\frac{1}{2}$ grammes, 48 grains or $\frac{1}{10}$ oz.
Granulated lead .... 3 " 90 " 1,440 " 3 "

Use no borax glass; instead :

Powdered silica ....$\frac{1}{20}$ A. T., 1$\frac{1}{4}$ grammes, 24 grains or $\frac{1}{20}$ oz.

4. Gray copper ores :

Ore................$\frac{1}{10}$ A. T., 2$\frac{1}{2}$ grammes, 48 grains or $\frac{1}{10}$ oz.
Granulated lead......2 " 60 " 960 " 2 "
Borax glass .... 300 mgrms., or 5 grains.

5. Sulphurets of iron :

Ore .... $\frac{1}{4}$ A. T., 7$\frac{1}{2}$ grammes, 120 grains or $\frac{1}{4}$ oz
Granulated lead.....2$\frac{1}{2}$ " 75 " 1,200 " 2$\frac{1}{2}$ "
Borax glass .... 200 mgrms., or 3 grains.

Litharge can be used to advantage in this as in other unroasted sulphurets.

Always weigh the litharge, so as to allow for its contained silver.

6. Oxide of iron :

Ore	⅙ A. T.,	5 grammes,	96 grains or	⅙ oz.
Granulated lead.......	1½ "	45 "	720 "	1½ "
Silica		1 "	or 15½ "	
Borax glass	350 mgrms., or 5½ grains.			

The silica to be mixed with the charge—it can be diminished as the percentage of silica in the ore increases.

7. Galena :

Ore..................	½ A. T.,	15 grammes,	240 grains or	½ oz.
Granulated lead.......	1½ "	45 "	720 "	1½ "
Borax glass	100 mgrms., or 1½ grains.			

Gentle heat. A nail in the scorifier aids in the desulphurization.

8. "Carbonate" ores :

Ore	⅙ A. T.,	5 grammes,	96 grains or	⅙ oz
Granulated lead	2 "	60 "	960 "	2 "
Borax glass...........	½ gramme or 7¼ grains.			

9 "Chloride" ores :

Ore..................	⅙ A. T.,	5 grammes,	96 grains or	⅙ oz.
Granulated lead.......	1⅙ "	33 "	576 "	1⅙ "
Borax glass	300 mgrms., or 5 grains.			

Use as low a heat as possible until the

charge has covered over, then heat more strongly to complete fusion.

10. Blende :

Ore ....⅛ A. T.,   5 grammes,   96 grains or ⅛ oz.
Granulated lead 3   "   90   "   1,440   "   3 "
Borax glass ... . 400 mgrms., or 6 grains.

Needs a good heat and care in its assay.

11. Arsenical and antimonial ores :

Ore ....⅕ A. T.,   5 grammes,   96 grains or ⅕ oz.
Granulated lead 4   "   120   "   1,920   "   4 "
Borax glass.   ... .. 1½ grammes or 23 grains.

This charge may have to be divided — sometimes requires several re-scorifications. A piece of charcoal laid over the scorifier, or some of it pulverized and dropped therein, often aids the fusion, as has been previously remarked.

12. Tellurides :

Ore   . ... 1/10 A. T., 2½ grammes, 48 grains or 1/10 oz.
Granulated lead..... 2   "   60   "   960   "   2 "
Litharge ....1/10 "   2½   "   48   "
Borax glass.. .... 250 mgrms., or 4 grains.

Sprinkle the litharge over the mixed charge. The buttons may need repeated scorifications with plenty of lead (20 to 1).

13. Native gold or silver, or *very rich* ores of any kind:

Ore............ ... $\frac{1}{10}$ A. T., 2$\frac{1}{2}$ grammes, 48 grains or $\frac{1}{10}$ oz.
Granulated lead ....1$\frac{1}{2}$ " 45 " 720 " 1$\frac{1}{2}$ "
Borax glass ... .250 mgrms., or 4 grains.

Ores, either those already described or any others, having a great quantity of lime or baryta, will require more borax glass than the quantities given. It may need to be as much as the ore taken — in such cases the charge may have to be divided, and several scorifications and re-scorifications made. Such large quantities of borax are best added at intervals, not at the beginning. Stirring will often aid. Give a strong heat. Make use of silica also.

The weights of borax glass are put down more to serve as indications than for preciseness' sake. It need not be weighed at all; after a time the assayer will learn to use it in *pinches.*

Study and experiment are necessary here. Ores will be found made up of several of those given separately above — new combinations of minerals are constantly coming to light.

The operator must then work over the particular ores he comes in contact with, until he learns them thoroughly.

The colors shown by the interiors of scorifiers are often characteristic of the ores tested, and in conjunction with the colors of the cupels after cupellation make valuable tests. The following descriptions may perhaps be of some value to the student in this connection :

COPPER.

Scorifier (See color plate).—A green, more or less deep, according to the percentage of this metal. Where the coating is somewhat thick, as on the edges of the scorifier, the color becomes a dark brown, but the prevailing tint is green.

Cupel.—Slate green, ranging to blackish green — prevailing tint very dark blackish green. There is frequently a *rose* coat on the

SCORIFIER
COLORS.

URANIUM.

CHROMIUM.

MANGANESE.

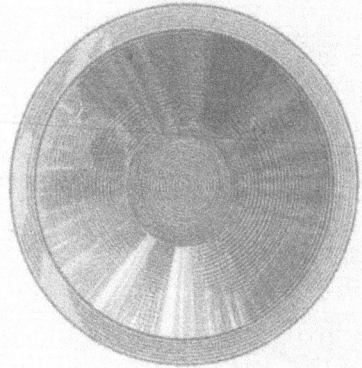

LEAD.

IRON.

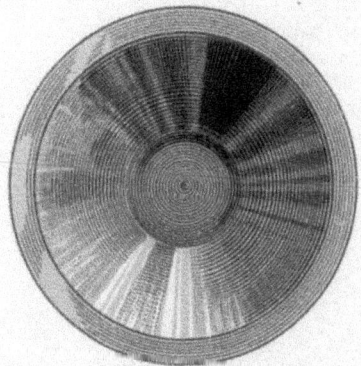

COPPER.

outside of cupels from ores rich in copper, which need not be mistaken for the rose color of oxide of silver. This rose tint meeting the slate green produces a purplish black or purplish green.

IRON.

Scorifier (See color plate).— With large amounts of iron the interior is black with a gray metallic lustre (this indicates a poor fusion), from which the color ranges down through a deep rich mahogany and varying shades of red-brown to a light yellow-brown. A red-brown is, however, always present, and is the prevailing color.

Cupel.— A brown tint, more or less decided.

LEAD.

Scorifier (See color plate).—Various shades of lemon yellow.

(Vanadium, cadmium, and bismuth give same shades.)

Cupel.— Bright, rich yellow to lighter shades.

(Small amounts of antimony, arsenic, bismuth, cadmium or zinc retained in the button also leave yellow markings on the cupel.)

MANGANESE.

Scorifier (See color plate).—From a purple-black to a light violet-brown or amethyst color.

Cupel— Blackish green to a lighter green, usually the latter. It is never so deep nor the cupel so thoroughly permeated by the color as is the case with copper ; the color also is different.

CHROMIUM.

Scorifier (See color plate).— Blood-red — thicker portions of the glaze a green, but prevailing tint is an orange or blood red.

Cupel.—Lemon-yellow, with reddish edges, surface also somewhat mottled with reddish-brown blotches — characteristic appearance.

NICKEL.

Scorifier.— A dirty, brownish yellow, and not very characteristic. A lead ore with a little iron would imitate it exactly.

Cupel.—Brown, almost identical with iron.

URANIUM.

Scorifier (See color plate).—A peculiar red.

Cupel.—Brown, as for iron.

COBALT.

Scorifier. — A beautiful blue ; toward the bottom it struggles with a green, which is probably due to the union of the yellow oxide of lead and the cobalt blue. This color will not be obtained from any ore, but it is probable that it has its effect in modifying other colors.

Cupel.—Brown, as for iron.

TELLURIUM.

Scorifier.—A yellowish color, with some red spots ; not very characteristic.

Cupel.—Yellow, with small green stains.

It must be borne in mind that combinations of the metals are liable to influence these colors and to produce mixed shades.

Also both the scorifier and cupel must be examined, not one alone, for the scorifier may reveal one metal, the cupel another. Thus an ore containing about equal parts of copper and iron gives a red-brown in the scorifier, indicating iron, while the green of the copper is almost entirely masked. On the other hand, the cupel is green-black, indicating copper, while no brown of the iron is visible; hence both tests prove both metals to be present in the original ore.

b. Scorification.

Place the scorifiers, by means of the scorifier tongs (page 100, fig. 50) in the middle and back of the muffle, which should be *decidedly hot*, close the door and augment the draft.

Then begins the first operation, the melting or fusion of the lead, due to intense heat and absence of oxygen, which takes from three to four minutes.

When the lead is liquid, open the door, thus admitting a current of air to supply oxygen, and which will also tend to diminish the heat somewhat.

Now, in the case of ores containing or retaining antimony, arsenic, sulphur, or zinc, a second operation, roasting, begins and continues till the greater proportion of the substances named have volatilized, the remainder of them going into the slag. ·

During this time the borax glass has melted and begun uniting with the gangue of the ore and with oxide of lead to form a slag which surrounds as a ring the molten lead.

As the scorification goes on, the melted lead grows smaller and smaller by oxidation and the volatilization of the greater part of the oxide formed, while the ring of slag gradually closes in and finally covers the lead, which is seen no more.

Finally increase the heat for a minute or two to fully liquefy the slag, which will finish the process of scorification.

Remove the scorifiers, and pour their contents into the cups of the scorification moulds (page 105, figs. 60 to 63), which should not be cold, covering each receptacle with its proper scorifier to retain its identification. (If *neces-*

sary, these scorifiers can be again employed for ores, etc.)

After having poured the charge, it will be well to let the slag and button remain in the cavity of the mould until they are stone cold before dumping them out, as otherwise there is danger of the lead adhering firmly to the slag.

Instead of pouring, the leads can be allowed to cool in their scorifiers, but no advantage is gained by this, and they take a longer time to cool.

In either case, however, when cold, detach the lead buttons from their slags, and hammer each button into a clean cube with flattened corners (fig. 128). Were the corners to be left sharp, they would injure the cupel when the button came to be dropped into it.

Fig. 128.

The weight of the button will vary according to the conditions ; the nature of the ore, the size of the charge, the heat of the furnace and the length of time the charge was allowed to remain in it, all exert an influence. A good weight is from twelve to sixteen grammes, which will make a cube of about one-half inch.

The button of lead is to be marked with some identifying number or letter with the point of a file or knife-blade.

The button should be perfectly malleable; if brittle it has probably retained antimony, arsenic, zinc or litharge, which can be gotten rid of by re-scorification. But with ores *very rich* in gold proceed with care, for the brittleness may be due to the gold itself, as beyond a certain limit gold takes away from the malleability of lead. If the button is large no extra lead need be added; if small an assay ton or two may be melted with it.

Again, the button may be very hard on hammering or show red in places, and perhaps on taking out of the scorification mould may have mossy copper on the bottom. In such cases the button must be re-scorified until no more copper is seen, or until it is very malleable. Plenty of lead must be used to alloy with the copper.

Since there is a greater loss of silver by cupellation than in scorification, very large buttons should be scorified down to a size suited to the cupels.

Examine the slag, and if it contains any globules of lead, hammer them flat, then place them on top of the main button, and cupel all together.

The slag should be vitreous or glassy, and of uniform character, its color depending upon the nature of the ore.

The scorifier should be perfectly smooth in its interior, that is, it should have no semi-fused lumps adhering thereto. Occasionally it may be corroded or eaten away, which does not necessarily injure the assay, unless the corrosion extends through the dish and allows its contents to flow out upon the floor of the muffle. In such a case (when of course the assay must be repeated) at once cover the floor of the muffle with dry sand or bone-ash, using the muffle shovel (fig. 58), and scrape out the mass adhering to it by means of the hoe or scraper (fig. 64). If this cleaning out of the muffle after an accident by spilling or leakage is not attended to, it leads to either one or both of two evils: first, the melted lead and borax attack the muffle and rapidly eat a hole through it; secondly, they stick to any scorifier or cupel placed in the muffle, making it almost impossible to move or remove either without breakage or loss of contents.

The corrosion of the scorifier is a good hint to add silica to similar ores, for usually it is the lack of this in the ore that causes the abstraction of silica from the scorifier, though there are times when a mixture of much lead and little ore is being scorified, that the lith-arge formed by the oxidation of the lead itself attacks the scorifier, and again, as in case of compounds rich in copper (a copper matte, for instance), the oxide of copper attacks the scorifier.

Sometimes in the process of scorification a crust forms over the surface of the charge and refuses to break. Such a crust is generally due to arsenical and antimonial ores present, and may often be destroyed by dropping in the scorifier some powdered charcoal wrapped in a wad of thin paper.

The oxidation can also be commenced by stirring the charge with a bent wire, until the lead is uncovered and begins to act. Withdraw the wire, break off the mixture adhering to the end and return it (the slag, etc.) to the scorifier, as it will probably carry some of the ore.

c. Cupellation.

This operation consists in oxidizing the lead of the lead buttons, the litharge formed by the heat being partly absorbed by the cupel and partly driven up the chimney, leaving the gold and silver together as a bead upon the surface of the cupel. Other metals that may have remained in small quantity from the previous operations, are also oxidized and so gotten rid of.

Take a good cupel (pages 114–117, fig. 79), in weight about one-third greater than that of the button that is to go in it, blow out any dust or impurities from the interior, mark on its sides in three or four places with ruddle or the point of a file, its appropriate number or letter, and with the aid of the cupel tongs or cupel shovel and hoe, place it in the muffle and there let it remain some four or five minutes that it may acquire the temperature of the furnace.

As can be inferred from the preceding paragraph, the size of the cupel depends upon the size of the lead button. And as mentioned under cupel-making, it is a good plan to have on hand cupels of various weights. It is stated that a good cupel will absorb its own weight of litharge, and furthermore, it is able to take a button heavier than its own weight, for a large amount of litharge (or oxide of lead) is driven off in fumes and consequently does not enter into the body of the cupel. But it is better to employ a cupel the weight of which is from one-fourth to one-third more than that of the button, for when a cupel becomes nearly saturated with litharge, the cupellation proceeds too slowly, when, on the contrary, it ought to be somewhat hastened, and cases occur that the cupellation ceases, though there may be at the bottom of the cupel enough unattacked bone-ash to absorb the remaining lead.

At other times an assayer may carelessly put altogether too large a button in a cupel, and therefore all the bone-ash in the cupel be saturated with litharge while there is yet melted lead above. This error may sometimes be rectified by putting a second cupel, red hot and inverted, under the soaked cupel, when the cupellation will proceed, though slowly. But it is better to reduce by scorification, in the first place, the excessively large button.

When the cupel or cupels have been in the muffle a few minutes, and consequently have become of the same temperature as the interior of the muffle, the lead button or buttons are to be placed in them, each one in its proper cupel, by means of the smaller curved tongs (page 100, fig. 48), and the muffle-door of the furnace closed, having previously, if necessary, placed a couple of pieces of coke or charcoal in the mouth of the muffle.

If the muffle has been of the proper temperature, in a minute's time or less, all the lead buttons will have quietly fused, and, on opening the muffle-door, each will be seen as a little lake of molten metal, from which arise fumes of oxide of lead.

The closing of the door at first is simply in order to melt the lead buttons, by the increased heat and absence of air.

It is very difficult to give in words directions for the proper conducting of this important step of cupellation. Experience is the best instructor.

In general, *do not have the furnace too hot.* This is not a matter of so much importance in the cupellation of the lead buttons from gold ores, but in those from rich silver ores it is such.

" The heat is too great when the cupels are whitish, and the metallic matter they contain can scarcely be seen, and when the fume is scarcely visible and rises rapidly to the arch of the muffle" (Mitchell), and *particularly* when the melted lead *bubbles.*

" The heat is not strong enough when the smoke is thick and heavy, falling in the muffle, and when the litharge can be seen not liquid enough to be absorbed, forming lumps and scales" (Mitchell), in short, to speak seemingly paradoxically, when the muffle and contents *look* cold.

An extremely high heat is bad, but a low heat is worse. " When the degree of heat is suitable the cupel is red, and the fused metal very luminous and clear" (Mitchell), and when scales of litharge are found in *small quantity* around the inner circumference of the cupel; in short, this " feathering " shows that the fire has not been too hot.

All this time, however, the buttons have been growing smaller and smaller, by oxidation and by volatilization and absorption of the oxide, changing from flat liquids to convex ones, and this reduction continues until we reach the point when the last of the lead leaves the bead. This is known as the "brightening," "flashing," "blicking," "coruscation," or "fulguration." As the button of gold, silver, and lead arrives near this stage it appears to revolve with great velocity, and rainbow colors succeed each other all over its surface. Finally a film passes over the bead, and then no more action is visible.

(With poor silver ores and ordinary gold ores the final bead is so small that it is difficult, if not impossible, to see the "blicking," but on beads from silver ores of any richness the brightening shows well that the operation of cupellation is concluded.)

Now move the cupel to the hottest place in the muffle, or increase the heat by closing the muffle door, that the last traces of lead may be driven off.   One source of error in silver assays is due to the assayer not getting rid of all his lead from the beads, but instead he weighs and reports it as silver.   Better err by under-reporting rather than over, so take the chances of volatilizing a little silver from the bead than to allow lead to remain with the silver.   A minute is generally sufficient to drive off the last lead, but with ores containing more gold than silver, let the cupel remain in the hot part three or four minutes, for there is no danger of losing any gold in that time.

Very rich ores betray themselves by a peculiar mottled appearance of the molten lead shortly after the cupellation begins.   The luminous blotches of litharge as they form string themselves out and cover the lead as with a network.   This is very characteristic, and once seen is again easily recognized. This mottling appears also with almost any buttons a little before the blicking ; in short, the richer the button the sooner it is observed.

Silver beads, on being suddenly brought from the hot interior of the muffle to the front where it is cooler, or out into the open air, sometimes "spit" or "blossom"— that is, the bead sprouts or vegetates, forming foliated protuberances all over its surface. This may occasion loss, as the spitting throws off particles of the silver; hence, guard against this as much as possible by moving the cupel by degrees to the front, and when at the mouth of the muffle cover with an inverted hot cupel. With beads weighing less than 30 milligrammes or thereabouts this need not be done, but above that weight proceed carefully.

If the assayer is running a number of assays, let him so arrange the cupels that those intended for buttons from poor silver ores or gold ores shall be in the center or hottest part of the muffle, while those for rich silver ores shall be in the fore part or cooler section. The reason for so doing is this: silver is sensibly volatile at a high heat, and the higher the temperature the greater the loss. On the other hand, the smaller the percentage of silver in a silver-lead, the less loss of this metal. By therefore placing the rich silver-lead in the cooler portions, the tendency is to decrease the loss by volatilization.   With any furnace, the heat of which cannot be instantly controlled, the muffle often becomes a little too hot for perfect cupellation.   When but few cupels are therein this does not matter much, since they can be slid to the front; but it is of importance when the muffle is so well filled that it becomes difficult or impossible to move any particular cupel or group of cupels to a cooler spot.   By now putting in the muffle a small *cold* scorifier or cupel, letting it rest on the edges of four of the cupels, the interior can be cooled down considerably.   Several scorifiers or cupels thus arranged have quite a lowering effect on the temperature, at least for a time.

When many cupels are being managed at once make a chart of their relative positions in the muffle, that there may be no " cases of mistaken identity" afterward, for with large buttons in small cupels the litharge often obliterates the ruddle marks.

If the furnace is too cold cupellation ceases, and the lead button is said to "freeze," forming a bunchy mass which undergoes no further action. A piece of charcoal laid upon the cupel, and additional heat applied, will sometimes finish the cupellation, or the button may be dug out of the old cupel, wrapped in a piece of lead-foil, and be re-cupelled in a new cupel. The result either way is none too accurate.

The final silver and gold bead from any cupellation should adhere with some tenacity to the cupel, have a bright, rounded surface, and appear frosted below.

d. *Weighing the Gold and Silver Bead.*

When cold detach the bead from its cupel, using the point of a knife-blade, and keeping a finger on the bead while so doing if the bead be small, for otherwise the exertion put forth to loosen the bead might easily snap it out of the cupel and past finding.

Lift the bead from the cupel by means of delicate pincers (p. 126), and cleanse from any adhering cupel dirt by rolling in the palm, by using a small stiff brush, or, if necessary, by flattening a little by means of a small steel hammer and anvil. If the bead be very small, fold it in three or four thicknesses of tissue paper, to prevent its flying away under the strokes of the hammer.

Weigh on the bullion scales in milli-grammes and fractions.

c. Parting.

The separation of gold and silver by dis-solving out the latter is designated by the term "parting."

The bead after weighing is flattened a little if it has not been so treated before. Now place in a little clean porcelain capsule or cru-cible (fig. 85), and fill about a quarter full with water (free from chlorine, see p. 156), and add four to six drops of concentrated nitric acid. No exact rule as to the amount of acid to add can be given, nor indeed is it necessary. But in general add drop by drop till it begins to "bite" the bead — that is, when the latter seems in violent motion and bubbles are thrown rapidly off. Instead of adding concentrated acid to water containing the bead, until it takes hold of the latter, the assayer may use a diluted acid of known strength. 16 parts of nitric acid of 41° Beaumé (specific gravity 1.41) with 30 parts of distilled water will make an acid of 21° Beaumé (specific gravity 1.16). This will do for ordinary small beads; for large ones, after having treated them with the above 1.16 acid, add some of 32° Beaumé (specific gravity 1.26), made by mixing 16 parts of the strong 41° acid with 10 parts of distilled water. Make these up in quantity and preserve in well stoppered bottles.

Now place the capsule on a sand-bath or wire triangle, and heat gently, not enough to cause the acid solution to boil. After a time no more action goes on. If there is no gold in the bead it will not blacken on adding the acid, and nothing will remain undissolved in the capsule; it will contain only the clear solution of nitrate of silver, formed by the silver dissolving in the acid.

In this case nothing further need be done than to wash out the contents of the capsule into a bottle containing silver residues.

But should one or more black specks be seen at the bottom of the capsule or floating about in the liquid, gold may or may not be present; at all events, these specks, *however small*, must be treated as though they were gold. Pour off the liquid above the black particles, first lightly tapping the capsule in order to cause the floating gold to settle to the bottom. If tapping will not either settle the gold or bring the particles together, try "churning"—that is, stir the contents of the capsule vigorously with a glass rod. In many cases, and unless the gold is in too fine a state, this will coagulate the gold, as it were — that is, bring the particles together into a spongy mass — when tapping will quickly settle it. It is best to pour into another clean porcelain dish, so that should the gold, by some mischance, go over with the outpouring solution, it may be recovered. Fill up the capsule with water, care being taken that no speck of gold is blown out of the capsule by the jet of water from the wash-bottle. This is to wash out the nitrate of silver from the gold. Tap the gold to the bottom, pour off the washings, and repeat the washing. If there is much gold, a third washing may be necessary. All this to insure complete removal of the silver nitrate. Finally drain off, wipe the capsule dry, remove, by means of filter paper (or clean blotting paper), any drops of water adhering to the interior of the capsule (being careful not to take away any of the gold), and heat *very gently* at first till all moisture has been driven off, then intensely for a minute or two. The gold has now changed in color from black to its normal yellow, and is very nearly pure, enough so for all practical purposes.   Let the capsule and contents cool.

f. Inquartation.

When a bead of gold and silver contains the gold in a greater proportion than about one-third of the silver, it possesses the power of resisting the solvent action of nitric acid. A certain amount of the silver may dissolve, according to the relative proportions of the two metals, but the larger part of it will remain so enveloped by the gold, that the strongest acid will not attack it.

Hence we resort to inquartation, or the operation of producing an alloy of gold and silver in such proportion that the latter metal may be extracted by nitric acid.

By the color of the bead the assayer can judge whether it needs to undergo this operation.   If it be of a moderately yellow color or a brighter yellow, it will probably need it.   But there can be no doubt of it if it refuses to be acted upon by the acid.

Remove it from the capsule, and dry. Weigh some thin and pure silver foil, in quantity about twice the weight of the bead. Wrap the latter in the foil, and place both in a cupel (or in a small hole bored in the back of the cupel), and fuse them well together in the flame of a blow-pipe. When cool, remove the now largely increased bead from the cupel, flatten and part as directed.

Instead of employing the blow-pipe, the bead and silver can be enfolded in some sheet-lead, and be re-cupelled in the usual manner. Indeed, if the original bead weighs more than ten milligrammes, it will be easier to alloy it by cupellation than by blow-piping, and a much better fusion be obtained.

g. Weighing the Gold Residue.

By means of a pointed slip of wood or sharp knife-blade, transfer the gold (which should be one scale or film) to the scale-pan of the bullion balance, and weigh with *exceeding care*, as usual in milligrammes and fractions. It often happens that the minute black pin-point of gold becomes too small to be weighed after the heating. It can then be reported only as a "trace" or "color."

h. Calculations.

By the use of the system of assay ton weights, the calculation of the gold and silver value of an ore becomes very simple. Two examples will show this very clearly:

EXAMPLE NO. 1.

Amount of ore taken......................	$\frac{1}{5}$ A. T.
Amount of test-lead used.................	$1\frac{1}{2}$ "
	MGRMS.
Weight of gold and silver bead...........	8.50
" silver in $1\frac{1}{2}$ A. T. lead..........	.25
True weight of gold and silver bead	8.25
Weight of gold in the bead...............	1.10
Weight of silver in the bead..............	7.15

$7.15 \times 5 = 35.75 = 35\frac{3}{4}$ milligrammes $= 35\frac{3}{4}$ ounces per ton of silver in the ore.

$1.10 \times 5 = 5.5 = 5\frac{1}{2}$ milligrammes $= 5\frac{1}{2}$ ounces per ton of gold in the ore.

VALUE OF THE ORE:

Gold—$5\frac{1}{2}$ ounces @ \$20 67 per oz............	\$113.68
Silver—$35\frac{3}{4}$ " " 1.29 "	46.11
Total value per ton.......................	\$159.79

EXAMPLE No. 2.

Amount of ore taken...................... $\frac{1}{2}$ A. T.

Amount of test-lead used .... I "

MGRMS.

Weight of gold and silver bead.............. 231.90

Weight of silver in test-lead*............... 0.00

True weight of gold and silver bead 231.90

Weight of gold, "faint trace"............... 0.00

Weigh of silver in the bead 231.90

$231.9 \times 2 = 463.8 = 463\frac{8}{10}$ milligrammes $= 463\frac{8}{10}$ ounces per ton of silver in the ore.

Value : $463.8 \times \$1.29 = \$597.30$ per ton.

CRUCIBLE PROCESS.

This process is much more complicated than the scorification, and, to use it successfully, we should know pretty thoroughly the nature of the re-agents employed, the kind and degree of their re-actions upon each other and upon the ore while in the crucible subjected to heat, and, finally, the characters of the various ores and their modes of behavior

---

* Not deducted. Read remarks on testing of granulated lead for silver, p. 173.

in the crucible.   Knowing all these things, we can decide upon such a modification of treatment as is best adapted to the particular ore in question.

The crucible process can be applied to any gold and silver ore, of whatever mineralogical nature, and whether rich or poor, but it has been found by experience to be best fitted for certain ores and classes, as is the scorification for certain others.

By some assayers the crucible process is reserved almost entirely for *low grade ores* (whether in gold or silver), and the scorification for *high grade ores*, and, for a general rule, this will do very well.   Another broad distinction is that of confining the crucible process to gold ores, and the scorification to ores of silver.   This latter rule is quite a safe one to follow, since the majority of gold-bearing ore is low grade, and hence is best assayed by the crucible process, which operates upon a larger quantity than the scorification.   Further, an ore of silver which is so poor as to give no results by the scorification process, is practically worthless, and needs no more testing by any method.

The following classification covers the ground, both generally and, in a measure, specifically:

CRUCIBLE PROCESS.	SCORIFICATION PROCESS.
1. Gold ores.	1. Silver ores.
2. Low grade gold ores.*	2. High grade silver ores, with the exception of the chloride and allied ores.
3. Low grade silver ores.	
4. "Chloride ores" (silver), including:	
Chlorides,	3. High grade gold ores.
Bromides,	4. Telluride ores.
Chloro-bromides,	5. Arsenical and antimonial ores.
Iodides.	6. Ores containing tin.
	7. Ores containing nickel or cobalt.

An outline of the crucible process is as follows :

The ore is mixed with lead in some form,

* For the purposes of assaying, a gold ore which runs over $5 to the ton may be considered high grade : below that figure, low grade.

fluxes, with or without some reducing, oxidizing, sulphurizing, or desulphurizing agent, transferred to a crucible and heated. The reduced lead absorbs the gold and silver, and settles with them to the bottom, while all impurities are fluxed (*i.e.*, slagged), matted, or volatilized. After removal from fire the lead is freed from slag, and the gold and silver separated from it by cupellation, as previously described.

To elaborate the above description somewhat, we may state, that, in the crucible process, as in the scorification, lead is added (save for lead ores) to extract the precious metals from the ore ; but, instead of using it in the metallic state and oxidizing the excess, we start with a compound which is almost invariably the oxide, litharge, and by means of a reducing agent, or by the reducing action of the ore itself (or by both combined), reduce enough lead to retain all the gold and silver, but which amount of lead shall not be too large to be manageable.

To secure a perfect crucible fusion, which means practically the separation of the gold and silver from everything else that may have been associated or combined with them in the ore (excepting, perhaps, some of the lead), we need to do several things. To get rid of the gangue (or earthy portion of the ore) we use fluxes, which convert the gangue into a slag. All metals present, save the precious ones, may be removed in various ways : By uniting them with sulphur into a matte ; by oxidizing them and fluxing the oxides into the slag ; by oxidizing them and volatilizing the oxides entirely out of the crucible ; or, in the one case of lead, by bringing either the whole, or such a proportion of it as may be wanted, down with that obtained from the litharge. Next, we may or may not need some reducing agent to reduce the requisite amount of lead ; or, on the contrary, an oxidizing agent may be necessary to oxidize and so remove sulphur or the excess of lead when an ore of this metal is being treated. Finally, a protecting cover of some easily fused inert substance will be wanted, although it is not indispensable.

The chief fluxes are litharge, the carbonates and bi-carbonates of potash and soda, borax, silica, and nitre. The reducing agents are charcoal, argol, cream of tartar, flour, or some other similarly carbonaceous substance, for it is the carbon of the reducing agent which removes the oxygen of the litharge, leaving metallic lead. The most commonly used oxidizing agent is nitre. Sulphur is the sulphurizing agent, and common salt the protecting cover. (The chemical composition and reactions of all the above are given in detail in the chapter on re-agents, which see.)

It will now be well, and in order, to undertake a little study of gold and silver ores, or of ores which are imagined to contain the precious metals, that we may know how to apply to them their proper treatment, which includes a knowledge of the principles of the important art of fluxing.

Every ore must belong in some one of the following divisions :

Metalliferous min- eral or minerals with *no gangue;* examples, p u r e galena, iron py- rites, or any "con- centrates."	Metalliferous min- eral or minerals with *a gangue;* examples, galena or iron pyrites in quartz.	Gangue matter with no *perceptible* met- alliferous mineral or minerals ; ex- amples, q u a r t z , fluor spar or ba- rytes.

The third division can, of course, include specimens that actually contain no metallifer- ous minerals, perceptible or imperceptible, as, for example, pure white quartz, while in others, as stated, they may be present so mi- nutely disseminated as to be invisible to the naked eye (or even to the eye assisted by a good magnifying glass), as is often true of samples carrying gold in that very fine condi- tion known as " flour gold."

There are also instances where that which is ordinarily a metalliferous mineral may act as a gangue ; for example, galena in spathic iron. Here the latter is a gangue, whereas at other times it may be found as a mineral

in a true gangue.   No confusion, however,
need arise, if we consider only the question
of the composition of the particular sample
lying before us.

The chief value of the above classification
lies in the fact that a knowledge of the class
to which an ore belongs aids us in its fluxing.

It would be almost impossible to write such
descriptions as would enable the student to
determine the mineralogical character of an
ore or to place it in its proper division above.
He must learn by experience, by observation,
and by an application of the information imparted by the standard authorities on mineralogy and blow-pipe analysis which are listed
in the appendix.   The few simple qualitative
tests given in the latter may aid him somewhat.

As regards the relative amounts of gangue
and mineral in an ore, I can give but one general, and rather indefinite, line of advice:
The heavier an ore, the greater the percentage of metalliferous mineral; the lighter an

ore, the greater the percentage of gangue. There are a few exceptions to this rule; the chief one commonly met with is barytes or heavy spar, a gangue which is three-fifths as heavy as galena, the chief ore of lead.

If an ore to be treated belongs to the first division — that is, if it possesses no gangue — we supply an artificial one, as it were, when we come to prepare the charge, by adding silica.

For purposes of assay treatment, we may consider under one heading those ores which appear to be all gangue, or made up of both gangue and mineral.

The nature of the gangue is important in determining the nature of the flux necessary to change it all into a slag. The gangue may be *acid, basic,* or *both acid and basic;* consequently, the simple rules for fluxing a gangue are almost self-evident. *An acid gangue requires a basic flux; a basic gangue requires an acid flux.* Now, what do we mean by an acid or a basic gangue? An acid gangue is simply one which acts as an acid, requiring a base to form a salt. A basic gangue, on the contrary, acts as a base, and therefore requires an acid to form a salt. As examples of these chemical facts, take metallic copper, which is a base. To convert it into a salt of copper we need an acid, which may be sulphuric, this making sulphate of copper (common blue vitriol), or silicic acid, forming silicate of copper (the mineral chrysocolla is the hydrated silicate of copper), or any other acids, forming corresponding salts. So in fluxing, an acid gangue, as silica, forms salts with a basic flux, as litharge or soda, producing a slag which is composed of the silicates of lead and soda, and a basic gangue, as lime, unites with acid fluxes, as silica or borax, producing a slag which is composed of the silicate and borate of lime. There results then the following classification :

ACID GANGUES.

1. Quartz, or other forms of _uncombined_ silica; as quartz c r y s t a l s, quartz rock, quartz-ite, sandstone, sand, etc.

2. Silicates, or silica _com-bined_ with some base; as clay, clay slates, mica, etc.

3. Rocks in which silica _predominates;_ as gran-ites, feldspars, por-phyry, etc.

As a rule, then, an acid gangue is silicious.

BASIC GANGUES.

1. Calc spar (carbonate of lime). Also l i m e-stones.

2. Heavy spar (barytes, or sulphate of baryta).

3. Fluor spar (fluoride of calcium).

4. All so-called earths; as alumina, and various combinations of lime, magnesia, b a r y t a, etc., _without_ silica.

5. Sparry iron, or carbon-ate of iron.

6. Various metallic o x -ides; as those of iron, m a n g a n e s e, etc., when in sufficient quantity to be con-sidered as gangues.

We can easily see now that the gangue of an ore may be both acid and basic, by its being made up of representatives of both classes. Theoretically, then, such a gangue should flux itself. Practically, however, we may find it necessary to help the fusion a little.

Try to ascertain the mineralogical character of the gangue. When this is known, consider only the element which is in excess of the others, and flux that ; the remaining ones will usually take care of themselves.

ACID FLUXES.	BASIC FLUXES.
1. Borax.	1. Litharge.
2. Silica.	2. Nitre.
3. Silicates ; as glass, and silicate of lead or lead-glass formed by the fusion of litharge and silica.	3. Carbonate of soda. Carbonate of potash. Bicarbonate of soda. Bicarbonate of potash.
Hence these flux the basic gangues enumerated.	Hence these flux the acid gangues enumerated.

The metalliferous minerals of an ore play a very important part in its treatment. They may exert either one of two very opposing actions, viz.: *reducing*, that is, taking away oxygen (from the litharge), or *oxidizing*, that is, giving up oxygen (to the reducing agent).

To explain a little more fully: Sulphur, arsenic, antimony, and zinc are the principal reducing elements of an ore. (If an ore contains none of these, it will not be at all reducing.) They all act in a similar manner to the carbon of a reducing agent, removing oxygen from the litharge, and so leaving metallic lead, themselves being converted into oxides. The various methods for eliminating these obnoxious substances will be considered further along.

An effect contrary to that produced by the elements mentioned is that caused by certain oxides and oxidized minerals. They are chiefly the oxides of iron, lead, copper, and manganese, *in their highest forms of oxidation*. Exposed to heat in a crucible, and surrounded by the reducing agent, they give up to it a portion of their oxygen, themselves being converted into their lower oxides and going into the slag.

The presence, then, of either reducing or oxidizing elements in an ore is made manifest by the size of the resulting lead button, the former tending to produce lead, the latter to keep it in the oxidized state by appropriating to themselves the carbon intended for reducing. Thus, some ores will be more or less reducing, bringing down more lead than is wanted, just enough or too little, or they may reduce no lead at all. On the other hand, they may be oxidizing, not only *not* bringing down any lead, but preventing the reducing elements of the ore (if present) and the reducing agent added from exerting their action. Hence it is important to know what an ore's action is.

It is advisable to reduce a certain portion of the litharge in order to obtain a button of a convenient size for cupellation. It should be large enough to extract all the gold and silver from the ore, yet not be so large as to be too long time cupelling, which may cause a loss in silver. A standard weight is 15 grammes for those who use the metric system, or 240 grains ($\frac{1}{2}$ ounce) for those accustomed to the grain weights. If the buttons obtained should chance to be somewhat larger or smaller, it will make no material difference; but ·it is well to use either weight given, and calculate ores and oxidizing and reducing agents to such a basis.

For the sake of reference, I next give lists of the principal ores and minerals whose direct or indirect action upon the litharge has just been discussed.

Metalliferous minerals which have a *reducing* action :

SULPHURETS.

1. Sulphuret of zinc (sphalerite, blende, zinc blende, " jack " or " black jack ").
2. Sulphuret of manganese (manganblende, alabandite).
3. Sulphuret of iron (pyrite, iron pyrites, mundic).
4. Sulphuret of iron (pyrrhotite, magnetic iron pyrites).
5. Sulphuret of iron with arsenic (arsenopyrite, mispickel, arsenical iron pyrites).

6. Sulphuret of copper (chalcocite, copper glance, vitreous copper).

7. Sulphuret of copper and iron (chalcopyrite, copper pyrites).

8. Sulphuret of copper and iron (bornite, erubescite, variegated copper pyrites, " horse-flesh ore ").

9. Sulphuret of copper and antimony with sulphurets of iron, zinc, silver, mercury, bismuth, arsenic, etc. (tetrahedrite, gray copper, fahlerz).

10. Sulphuret of antimony (stibnite, gray antimony).

11. Sulphuret of lead (galenite, galena, "pyrites of lead," "mineral").

12. Sulphuret of silver (argentite, silver glance).

The above list enumerates the chief sulphurets which are likely to be assayed for gold and silver, though there are many other intermediate and mixed sulphurets, for descriptions of which consult the mineralogies. It can easily be remembered, however, that *any* sulphuret, or mixture of sulphurets, is reducing.

To the above add the various arsenides, antimonides, bismuthides, selenides, etc., of more or less rarity, and more commonly ores containing graphite or plumbago.

Metalliferous minerals which have an *oxidizing* action :

1. Red oxide of iron (" decomposed iron ore ").
2. Red oxide of lead (minium).
3. Black oxide of copper (tenorite, melaconite).
4. Black oxide of manganese (pyrolusite).
   (More rarely the chromates).

Since an oxidized ore commonly results from the decomposition of a sulphuretted one, it may be very possible to find an ore not entirely decomposed or oxidized, and therefore both reducing and oxidizing. As examples, one often meets with a specimen carrying sulphurets and oxides of iron, or copper pyrites and black oxide of copper intermingled, or galena and carbonate of lead.

A preliminary examination of an ore may then be necessary to determine its character and subsequent mode of treatment. An expert mineralogist, metallurgist, miner, or assayer can tell almost at a single glance the nature of an ore as to its constituents and oxidizing or reducing powers, but the beginner will find it somewhat difficult at first. Here then again comes in the opportunity to make blow-pipe and qualitative examinations.

To ascertain the presence or absence of sulphurets, try the following very simple test: Powder a little of the sample and heat it in an iron spoon over a strong flame or on top of a good fire, and note its behavior. If, shortly after it is heated, star-like sparks are quickly thrown off (particularly noticeable when the powder is stirred with a wire), then the mass begins to glow around the edges (resembling a charcoal fire), while closely above it hover small blue flames, and finally the entire mass becomes red-hot, with fumes and an odor as of a burning match is perceived, then sulphurets of iron or copper or both are present, and so the ore will be decidedly reducing. If, in addition, an onion or garlic-like odor and whitish fumes are noticed, arsenic (probably as arsenical iron pyrites) is also present. Antimony and zinc give white fumes with no odors. Blende and galena are not so liable to glow and to scintillate as are the iron and copper sulphurets, but the smell of a burning match should be perceptible upon heating them or any other ore which contains a fairly large percentage of sulphurets. It will also be noticed that after the mass has been thoroughly heated and allowed to cool, it will have lost its metallic shimmer, and become of a dull, dead color, indicating oxidation.

A simple chemical test for sulphurets may easily and quickly be tried. Place a small portion of the finely ground ore in a test-tube, pour in a little water, shake, add a few drops of hydrochloric acid, and warm gently. If the smell peculiar to rotten eggs (sulphu-retted hydrogen) is now recognizable, sulphurets are certainly in the ore.

Finally, when it is suspected that not a very large quantity of any sulphuret is in the sample, try the following plan, which is much more delicate than either of the preceding : Fuse a little of the ore with bicarbonate of soda and borax on charcoal, by means of the blow-pipe. When the mass has fused remove it, by means of a knife-blade, to the surface of a bright silver coin, add a drop of water, and work the paste thus formed for a short time. The same gas above spoken of (sulphuretted hydrogen) is given off from the sulphide of sodium, which has been made by the union of the sulphur of the ore and the soda, and blackens or browns the coin, according to the amount of sulphur in the original ore.

A very close guess as to the *kind* of sulphuret contained in an ore, or of which it is mostly composed, can frequently be made (depending on the purity of the sulphuret) by noting the color and character of the pulverized sample, then heating some *quickly* in a roasting dish, observing its action under heat, finally examining the heated ore, after it has cooled. Compare it with the following table :

| KIND OF SULPHURET. | BEFORE HEATING. | | DURING |
| | Color of Ore. | Character. | Fumes. |
			Character.
1. Blende	Brown, shading to green, red, and yellow.	Shining.	Very slight ....
2. Manganblende ..	Green, green-ish-gray.	Dull	Slight.........
3. Pyrite..........	Greenish or brownish black.	Shining, not magnetic.	Slight.. ....
4. Pyrrhotite .. .	Gray-black	Shining, is magnetic.	Slight.........
5. Arsenopyrite ...	Gray-black	Shining....	White, very thick.
6. Chalcocite	Gray	Shining	Slight.. ......
7. Chalcopyrite	Green-black	Shining......	Some fumes ...
8. Bornite	Dark-green	Shining...	Some fumes....
9. Tetrahedrite	Dark-gray	Shining....	Much.. . ....
10. Stibnite	Lead-gray	Shining...........	White fumes...
11. Galenite	Gray	Very decided shine.	Fumes con-siderably.
12. Argentite	Gray-black, pink and brown shades.	Very little shine ...	Some ....

(Each sulphuret will give a little odor of sulphur, but

Heating.		After Heating (when cold).		
Fumes. Odor.	Other Characteristics.	Color of Ore.	Character.	Composition.
None	Glows, yellow-green when hot.	Buff or yellowish.	Dead, dull	Oxide of zinc.
None	Glows..	Brown........	Dead, dull	Brown oxide of manganese.
None	Glows......	Red to black —many shades.	Dead, dull	Red oxide of iron.
None	Glows.	Red to black — many shades.	Dead, dull	Red oxide of iron.
Like garlic....	Glows, swells.	Red to black — many shades.	Dead, dull ...	Red oxide of iron.
None	Glows, fuses.	Gray	Shining, powder red.	Oxide of copper.
None	Glows, fuses	Gray	Shining, powder gray.	Oxides of iron and copper.
None	Glows.... .	Gray	Shining, powder brown.	Oxides of iron and copper.
Garlic odor sometimes.	Glows, swells, fuses.	Gray	Shining	Mixed oxides.
None	Swells, melts, boils.	Gray-green	A thin film of shining mass.	Oxide of antimony.
None	Fuses	Yellow-green ..	Shining	Oxide of lead.
None	Fuses	Gray	Somewhat shining.	Oxide of silver somewhat reduced.

it is not to be confounded with the odor of arsenic.)

An approximate determination of the percentage of "sulphurets" in an ore can be made by mixing 10 grammes or 100 grains of the ore with as much powdered silica and twice as much borax glass, placing the mixture in a suitably sized crucible, topping with salt, covering, and heating pretty strongly for fifteen minutes. Remove, cool, break out button as a "matte," weigh, and multiply weight by ten, giving percentage.

The blow-pipe tests for antimony and zinc are not satisfactory, especially in the hands of a beginner; hence, I have given the following analytical scheme, which, although it seems complicated, is really simple. It can be applied to the most complex ores; for the more simple ones some of the steps can be omitted, as the student will learn by experimenting with it on the various ores and minerals. To practice it, make a mixture of iron and copper pyrites, blende, galena, arsenopyrite, sulphuret of antimony, and black oxide of manganese. The test can be made in an hour's time.

Make a mixture, in equal parts, of hydrochloric, nitric and sulphuric acids. Place this in a beaker, and into it drop some of the powdered ore, and heat; then *add water*, and filter. Do not wash.

Residue and Precipitates.		*Filtrate.*	
Sand, silica, clay; mercury, silver, and lead as chlorides; lead, calcium, barium, and strontium as sulphates; antimony as oxychloride. Wash thoroughly with water while on the filter.		Add sulphuretted hydrogen gas, and filter from precipitate formed.	
Filtrate.	*Res. and Pre.*	*Precipitate.*	*Filtrate.*
Not needed, reject.	Remove from filter, boil with solution of tartaric acid, and filter.	Mixture of various sulphurets; reject.	Boil off excess of sulphuretted hydrogen, add caustic soda in excess, boil and filter.
Res.	*Filtrate.*	*Precipitate.*	*Filtrate.*
As first given; reject.	Run sulphuretted hydrogen gas through it, or add water saturated with this gas; an orange-colored precipitate indicates antimony.	Iron, chromium, aluminium, and manganese as hydrated sesquioxides.	Add acetic acid in excess, and treat with sulphuretted hydrogen; a white precipitate indicates the presence of zinc.

It is difficult to give tests for the highly oxidized ores listed. Dull, dead ores in general are likely to be in an oxidized condition, though not necessarily in the highest state of oxidation. If the ore is red and does not

answer to any test for copper, it will probably be the red oxide of iron (possibly of lead).
If it is black, apply the simple tests for copper and manganese given in the appendix.

But even without a knowledge of the ore,
gained by observation and experience, or by
applying tests, its reducing or oxidizing power
can be determined by certain preliminary and
arbitrary assays.

If it is suspected that the ore is reducing,
prepare the following

PRELIMINARY CHARGE TO DETERMINE REDUCING POWER.

*A. T. Weights.**	*Gramme Weights.**	*Grain Weights.**
Litharge ... $1\frac{1}{2}$ A T.	45 grammes	720 grains.
Ore........ $\frac{1}{20}$ "	$1\frac{1}{2}$ "	24 "
Salt cover.		

(The above weights of ores are each $\frac{1}{20}$ of
the respective standard weights taken for the
crucible process.)

Weigh first the litharge and brush it on to

* These weights do not correspond exactly with one another
(although sufficiently so), nor is it demanded that they should. They
simply represent the most appropriate quantities for the purpose in
the several systems of weights.   The student can take his choice.

a clean sheet of black glazed paper, then weigh very carefully the finely pulverized ore, sampling it as usual; brush it on top of the litharge, and mix them thoroughly (indicated by the uniform color and character of the resulting compound) with a large steel spatula. Pour the charge into a small sand crucible, which it should not more than two-thirds fill (size "W" of the Battersea make, 3 inches deep by $2\frac{1}{2}$ inches across, will do nicely); tap gently till contents are level, sprinkle some dry salt over the glazed paper, stir it around thereon by half turning over the sides of the paper, finally pour it on top of the charge — in this manner any of the charge left adhering to the paper is "dry washed" into the crucible. There should be about $\frac{1}{2}$ inch salt on top. There will then remain an inch or so space between top of crucible and the salt, to allow for the expansion of the charge when fusing.

Have the fire quite hot, place in it the crucible, covered, bank around it with the hot coals, and heat quickly till contents are in quiet fusion, which requires fifteen to twenty minutes.

When satisfied that the charge is well fused, remove cover, lift out the glowing crucible by means of the long-handled crucible tongs (figs. 44 to 48), and tap it gently on a brick three or four times (in order to gather into one button any little pellets of molten lead that may be scattered throughout the fused mass), then cover, and let stand till cold.

When stone-cold, break the crucible, detach from all adhering slag the lead button (if there is any), hammer into shape as usual, and weigh it.

We shall have one of two alternatives :

        1st. A lead button.

        2d. No lead button.

1st Case.   A lead button is obtained.

This result shows that the ore has a reducing action — the weight of the button will indicate its power.

There will be then one of three results :

A. The lead button has a weight-less than one-half that of the ore.

B. The lead button has a weight about equal to one-half that of the ore.

C. The lead button has a weight greater than one-half that of the ore.

I take each supposition in turn.

A. The lead button has a weight less than one-half that of the ore.

Rule. — Multiply the weight of the lead button by 20 (to ascertain what the standard weights of 1 A.T., 30 grammes or 480 grains would produce), and subtract the product from the standard weight of button desired (15 grammes or 240 grains). The result is the weight of lead which must be reduced by some reducing agent. Divide this by the reducing power of the reducing agent to be used, and the quotient will be the weight of the reducing agent necessary to be added in the regular assay.

Examples. — Suppose the $\frac{1}{20}$ A. T. or $1\frac{1}{2}$ grammes of ore produced a lead button weighing $\frac{1}{4}$ gramme, then 1 A. T. or 30 grammes would produce $\frac{1}{4} \times 20 = 5$ grammes, $15 - 5 = 10$ so that the lead button lacked 10 grammes of the standard weight. We have therefore that amount to be reduced by an added reducing agent. Taking charcoal as 25 and dividing that into the 10 gives us $\frac{2}{5}$ gramme as the weight of charcoal to be employed.

In a similar manner would we proceed when grain weights are used. If 24 grains of ore reduced 4 grains of lead, then $4 \times 20 = 80$; $240 - 80 = 160$; $160 \div 25 = 6\frac{2}{5} = $ grains of charcoal to be used.

Should the particular sample of charcoal employed possess a greater or lesser reducing power than 25, or should argol (reducing power about $7\frac{1}{2}$), or flour (reducing power about 15), or any other reducing agent be made use of, then simply substitute in the above calculations the appropriate figures.

B. The lead button has a weight about equal to one-half that of the ore.

This case is very simple. The quantity of

ore taken for the regular assay will, of course, as in the preliminary assay, reduce half its own weight of lead (giving buttons of about the right size), hence there is needed no additional reducing action.

C. The lead button has a weight greater than one-half that of the ore.

Here we have an instance of too great a reducing action, hence the excess of lead must be oxidized away, or, to put it more correctly, we must supply enough oxygen to satisfy that proportion of the reducing elements of the ore which would reduce the excess of lead, and then they will leave the oxygen of the litharge alone. The oxygen-supplying medium is nitre.

Rule. — Multiply the weight of the lead button by 20, and subtract from the product the standard weight of button desired. The result is the excess of lead which is to be retained in the form of litharge by use of nitre. Divide this by the oxidizing power of the nitre* used, and the quotient will be the weight of the latter to be employed.

Examples. — Suppose the $\frac{1}{20}$ A. T. or the $1\frac{1}{2}$ grammes of ore produced a lead button weighing 3.15 grammes, then 1 A. T. or 30 grammes would produce $3.15 \times 20 = 63$ grammes, $63 - 15 = 48 =$ weight of lead to be kept oxidized $48 \div 4 = 12 =$ grammes of nitre to use.

If the grain weights are used, and the 24 grains of ore have reduced say 50 grains of lead, then $50 \times 20 = 1,000$; $1,000 - 240 = 760$, $760 \div 4 = 190 =$ grains of nitre to keep oxidized the 760 grains of unnecessary lead.

The various sulphurets possess varying reducing powers ; hence, by starting with a knowledge of the reducing power of each kind of sulphuret, and estimating, by the eye, the proportion or percentage of whatever kind may be in the ore in question, we may approximate pretty closely to the amount of nitre to add in order to oxidize all the sulphur but that which we want present to bring down

* One part of nitre oxidizes four parts (very nearly) of lead.

the proper weight of lead, and so can avoid making a preliminary assay. The adjoining table will be found useful for reference :

KIND OF SULPHURET.	Parts of litharge required to completely oxidize 1 part of the sulphuret.	Parts of metallic lead reduced by 1 part of the sulphuret.	Percentage of the sulphuret which with charges of 1 A.T., 30 grms. or 480 grains, will reduce a lead button of about 15 grammes or 240 grains.	Parts of nitre required to completely oxidize 1 part of the sulphuret.
1. Zinc blende ...	25	6 5	7.7	
2. Manganblende .	30	6.7	7.5	
3. Iron pyrites ...	50	8.6	5.8	2.5
4. Arsenopyrite...	40	7.3	6.5	
5. Copper pyrites .	30	7.2	7	
6. Copper glance .	25	3.8	13	
7. Gray copper . .	35	6	8	
8. Gray antimony .	25	5.7	9	
9. Galena	1 3	2.8*	18	$\frac{2}{3}$

* All the lead of the galena and litharge.

Should the assayer decide to *roast* the ore, of course it will not be necessary to make a preliminary assay to determine its reducing power, as the roasting will eliminate the reducing elements. But it should be remem-

bered that the roasting converts the sulphurets of copper, iron, manganese, and lead, and
their lower oxides, into the higher oxides of
the same metals. It will, therefore, make
their ores highly oxidizing, requiring a greater
amount of reducing agent in the actual assays.
and, perhaps, necessitate preliminary assays
to determine their oxidizing powers, as shown
in the succeeding paragraphs.

2d Case. No lead button is obtained.

This result shows conclusively that the ore
has no reducing action. It may also have no
oxidizing action ; but, on the other hand, it
may. In case it has no oxidizing action, then,
for the regular assay, take such an amount of
whatever reducing agent is used as will produce the proper weight lead button.

Thus :

Grammes or Grains		*Of ordinary*	Will reduce
$2\frac{1}{3}$	38	Cream of tartar	a lead button
$1\frac{3}{4}$	28	Argol	of about 15
1	16	Flour	grammes or
$\frac{1}{2}$	8	Charcoal	240 grains.

To determine the exact oxidizing power of the ore, prepare and run the following

PRELIMINARY CHARGE TO DETERMINE OXIDIZING POWER.

	A. T. Weights.	*Gramme Weights.*	*Grain Weights.*
Litharge..	1 A.T.	30 grammes.	480 grains.
Ore	$\frac{1}{20}$ "	$1\frac{1}{2}$ "	24 "
Charcoal..	.050 grms.	.050 "	$\frac{3}{4}$ "
Salt cover.			

Use a " W " crucible, treat in the usual manner, and weigh resulting button.

As in the tests for reducing powers, there will be either :

      1st.   A lead button.

      2d.   No lead button.

1st Case.   *A lead button is obtained.*

The weight of the lead button will, of course, vary, giving rise to one of three results :

A. The lead button has a weight less than one-half that of the ore.

B. The lead button has a weight about equal to one-half that of the ore.

C. The lead button has a weight greater than one-half that of the ore.

To proceed in regular order.

A. The lead button has a weight less than one-half that of the ore.

Rule.— Multiply the weight of the lead button by 20 (to bring the calculations up to the standard weights of 1 A. T., 30 grammes or 480 grains), and subtract the product from 15 or 240. The difference is a part of the amount of lead kept oxidized by the standard weight of ore employed, and this action is to be neutralized by an extra amount of charcoal. Divide this difference by the reducing power of the charcoal used, and the quotient will be the weight of charcoal needed.

Examples. — The figures given were derived from assays on an ore which was composed of about one-half silica and one-half red oxide of iron. The charcoal used had a reducing power of 23.4 — that is, one part of the charcoal reduced 23.4 parts metallic lead from the litharge.

The charge given was run twice, and buttons weighing 0.30 grammes obtained. 0.30×
20=6.00; 15.00—6.00=9.00= grammes of lead
kept as oxide by the action of the ore. 9÷
23.4=0.384= grammes of charcoal to neutralize this action.   Hence,

	Grms. Charcoal.			*Grms. Lead.*
.050×20	=	1.000	to reduce	6.00
9÷23.4	=	0.384	" "	9.00
Or a total of ..		1.384	" "	15.00

Had there been no oxidizing ore present,
the 1 gramme charcoal would have reduced a
button of 23.4 grammes, as stated.   It actually reduced but 6 grammes ; hence, 23.4—6=
=18.4= grammes of lead kept oxidized by the
ore, but, as seen, we cared to reclaim only
9 grammes of this 18.4 grammes — the remainder can be left oxidized.

About the same calculations in the grain
system would be as follows : 4.62 grains =
weight of lead button reduced by ¾ grain
charcoal ; 4.62×20=92.4; 240—92.4=147.6=
deficiency in grains of lead of a 240 grain but-

ton ; $147.6 \div 23.4 = 6.3 =$ grains of charcoal to
reduce that deficiency. Hence,

	Grains Charcoal.			*Grains Lead.*
¾ or $\quad$ 0.75 × 20 $\quad =$	10.5	to reduce		92.4
147.6 $\div$ 23.4=	6.3	"	"	147.6
Or a total of......	16.8	"	"	240.0

*B. The lead button has a weight about equal
to one-half that of the ore.*

In this case simply multiply the charcoal
and ore twenty times for the regular charge.
For example, an ore with .050 grammes charcoal as usual gave me a lead button of 0.765
grammes ; hence for the regular assay I took
1 gramme charcoal to 1 A. T. or 30 grammes
ore to obtain a 15 gramme (15.30) button.

In other words, the ore oxidizes such a proportion of the charcoal that the remainder of
the latter is just sufficient to bring down a
button of the right size.

*C. The lead button has a weight greater
than one-half that of the ore.*

Here we have an example of too slight an
oxidizing action, which allows of too great a

reducing action on the part of the charcoal, resulting in too large a button. There is then needed more oxidizing action, which is effected by adding an oxidizing agent, or, which amounts to the same thing, by lessening the quantity of charcoal.

Rule.— Multiply the weight of the lead button by 20, and divide the product by 15 or 240. The quotient multiplied by the original amount gives the right quantity of charcoal to reduce a 15 gramme or 240 grain button.

Examples.— Imagine that the charge on a certain ore gave a button of 1.10 grammes. This multiplied by 20 is 22 grammes, which is too large a button ; hence, better diminish the quantity of charcoal. If 1 gramme charcoal with this particular ore brings down a 22 gramme button (instead of 23.4, the ore oxidizing the difference of 1.4 grammes), then $15 \div 22 = 0.68$; $0.68 \times 1 = 0.68$ grammes, the amount which will reduce a 15 gramme button.

In grains, let the lead button weigh 17 ; 17

grains × 20 = 340 grains (too large a button) ;
240÷340=0.7; 0.75×0.7=0.525, or say ½ grain.

   2d Case.   No lead button is obtained.

This would show that the ore is extremely
oxidizing, and to learn its precise strength
necessitates the repeating of the preliminary
with double, treble, etc., the quantity of charcoal, or till a button finally *is* obtained.   But,
in general, it will be safe to use twice the
weight of charcoal given, or for the regular
assay 2 grammes or 31 grains where the preliminary produces no lead or a very minute
button.

We have previously spoken of certain reducing elements in ores—viz.: sulphur, arsenic,
antimony, and zinc.   Aside from the reducing
action which they exert upon the litharge,
tending to produce too much lead, their presence in an ore is apt to affect the accuracy of
the crucible assay.   Thus the sulphurets may
combine with oxygen, forming oxysulphurets,
which possess the undesirable property of
taking silver with them into the slag, or, if

they remain undecomposed, they may easily retain some gold. Arseniates and antimoniates, whether existing naturally in the ore, or whether formed by oxidation during the fusion, are also liable to keep silver away from the lead button. Blende or other zinc ores carrying silver may volatilize this metal or retain it in the slag.

The objectionable elements mentioned may be gotten rid of in two ways : by oxidation in the crucible during the fusion, or by a preliminary and separate operation known as roasting. If the sulphur, etc., are in small quantities, the fluxes, litharge and soda, will oxidize them into the slag. If in large quantities, nitre must be used. It is yet a disputed question among assayers whether to roast an ore for crucible assay or not. Much can be said with truth on each side. If roasting is not performed we save the trouble of an extra operation, and the danger of loss which extra steps are always likely to cause, including the possibility of volatilizing silver by too great a heat in roasting. On the other hand, if roasting is not performed, and nitre is not used, there are the various chances of loss of precious metal already described. If nitre is used, there is danger of mechanical loss by the foaming of the charge, which may overflow the crucible or leave lead sticking to the sides, high up; or if too much nitre be taken, the litharge may not be reduced, so care must be exercised when using it. My own experience with beginners is that they are more successful when they previously roast refractory ores, than when they omit this step. It certainly brings the ore into a much more workable condition, as the oxides of iron and copper which are usually left are easily treated. With any particular ore the safest way to proceed is to run duplicate or triplicate assays on both the roasted and unroasted sample, and then to adopt or dispense with the roasting accordingly as the results are richer with or without it. For the benefit, then, of learners, for whom this book is written, and not for ad-

vanced assayers, I give herewith a very careful description of the manner of performing
this important operation.

Roasting.—Weigh the ore carefully, sampling as usual. Next transfer it to a sheet of
black glazed paper, and mix with it about its
own bulk of fine charcoal. The latter is usually recommended only when arseniates and
antimoniates are present or likely to be formed
by a plain roasting ; but I consider it advisable
to use it every time, for one cannot always
tell when small quantities of arsenic and antimony are in the ore, besides which, the charcoal aids in expelling the sulphur, and in indicating the termination of the roasting. At
the worst, it can do no possible harm. If the
ore should contain much sulphide of lead
(galena) or sulphide of antimony (antimony
glance), both of which are quite easily fusible,
add to the ore and charcoal on the paper
some fine sand or precipitated silica, and mix
all well together, for without this addition the
minerals while roasting would soon fuse, cake

together, and adhere to the dish or pan, thus ruining the assay. The proportion of silica to use depends upon the percentages of the minerals named, it increasing as they increase — roughly speaking, it can be employed bulk for bulk.

In case the ore is a sulphide or mixture of sulphides (of whatever metal or metals), with very little or no gangue, it is best to add silica, to the amount of at least three-fourths of the weight of ore.

The contents of the glazed paper are now brushed into either a frying-pan (page 131), or a clay roasting dish (page 112), accordingly as an open fire or the muffle is to be used. The frying-pan should be protected by a coating of dry chalk or ruddle (or a water paint of either can be used and the pan dried), or plumbago. The roasting dish may be similarly protected, but it is not so necessary. If the standard weights of 1 A. T., 30 grammes, or 480 grains is used, the ore mixture can be roasted in the largest roasting dish which will go in the muffle ; but above those weights use the frying-pan, or else roast several charges in the dishes, and unite them for fusion.

Place the roasting dish, with contents, in the forward part of the muffle, before the latter has reached a dull red heat. The ore is to be continually stirred with a stout wire, having a loop at the end at right angles to the wire, ⎯⎯⎯⎯⎯△. In a little while, minute sparks will be thrown off, and the ore will begin to glow in places like burning charcoal. Stirring should be continued till the glowing ceases (by which time there will be little danger of fusion), and the whole seems of one color, and yielding to the stirrer like dry sand. The dish can now be moved back to the hottest part of the muffle, and left unstirred for some little time. When, on bringing out to the open air and stirring, the ore gives off neither fumes nor odor, the operation is finished, the ore being now "sweet." Let cool, and examine.

The roasted ore should be of a dull, dead, earthy color (usually some shade of red or black), having no metallic lustre, containing no large, hard lumps, and having no portions adhering to the dish. See that all the charcoal is burnt out. Should the ore contain many lumps, grind in a mortar, mix with charcoal, return to the dish, and repeat the roasting.

If the assayer is too busy to spare the time needed for stirring, let him take the dish, and spread its contents out thinly, or ridge the ore from the center to the edges, so as to increase the surface exposed, and place it in the muffle, and warm it gently for a time; then increase the heat for about the same length of time; finally heat quite hot. Take out dish, cool, grind finely, return to dish, and roast again. Repeat the roasting two, three, or more times if necessary. All this will occupy time, but not the direct attention of the assayer.

The roasting in the pan is done similarly.

Whether open-air pan roasting, or dish in muffle roasting is performed, guard against too high a heat at first, as that may cause fusion, or too sudden a delivery of the volatile metals mentioned, which may carry off silver, or even too great heat at any stage of the roasting, tending to volatilize silver or gold from some of their combinations.

The roasting with charcoal is supposed to decompose all sulphates, arseniates, and antimoniates, and to expel them; but where the ore was or contained copper pyrites, a certain quantity of sulphate may remain unchanged. Hence, with ores of this nature, mix the roasted ore with from one-fourth to one-half its weight of fine and *dry* carbonate of ammonia. Return to dish, cover with an inverted roasting dish, and place in a moderately warm part of the muffle till no odor of ammonia can be perceived. The sulphate of copper is converted into sulphate of ammonia, which, being very volatile, is quickly driven off.

If the laboratory is provided with gas, the

ore can be roasted in the usual roasting dish over the Fletcher burners or the Fletcher roasting furnace, illustrated on page 74, which will work very nicely, as the temperature can be regulated to any degree.

Methods of the crucible assay.—There are two general methods or systems of crucible assays in use in this country. The first, more ancient, better known, and more commonly used one employs an *excess* of litharge, and we can therefore consistently call it the *litharge crucible process.* Mr. C. H. Aaron, in his very valuable little work on assaying, designates it as the first system, while another method, which he is the first to describe, he calls the second system. In justice to him, I prefer to call it *Aaron's crucible process.*

The distinction between the two systems is this: The first mentioned uses litharge for two purposes, to furnish lead enough for a lead button to retain the gold and silver, and to aid in fluxing the ore, hence an excess of the litharge is employed. The second uses just enough litharge to provide lead for the lead button.

In the first, the excess of lead not only slags off the gangue, but it oxidizes all the base metals, save lead, and takes them also into the slag. In the second, the gangue is slagged by soda, borax, or silica, while the base metals are either volatilized, united with sulphur into a matte, or combined with iron.

Any ore may be assayed by either of these processes, but each is better suited to certain classes.

Mr. Aaron argues that while an assay by the litharge process "is quickly made, and generally gives accurate results," yet "it has the disadvantage of requiring considerable modification for the various ores, as to the fluxes proper, and to the reducers or oxidizers by which the production of lead is controlled. Sometimes a preliminary assay is necessary."

For his process he claims the following advantages: "The right quantity of lead may nearly always be got at once, for, al-

though any lead which the ore may contain
will inevitably come down together with that
from the litharge used, yet this can be allowed
for by reducing the quantity of litharge, or
omitting it.   As litharge yields ninety-three
per cent of lead, it is not difficult to make
the adjustment nearly enough.   Galena contains eighty-six per cent of lead ; hence, if the
ore is nearly pure galena, but little litharge is
needed.   The method requires but slight
modification for different ores, and may with
little disadvantage be made universal.   The
button is never much contaminated by copper, as it often is in the other system, unless
a very large proportion of litharge is used,
which is disadvantageous in some ways.   The
crucible is but little attacked, and the assay is
not liable to boil over.   The method is especially useful for ores carrying much galena
or other sulphuret, and when copper in any
form is present."   An additional advantage,
resulting from the above, as remarked by Mr.

Aaron elsewhere, is the saving in litharge and crucibles.

My own experience with Mr. Aaron's process shows that it takes considerably more time than does the litharge process, and that beginners find it more difficult to operate, and to get satisfactory results.

The student can take his choice of either process, apply it to any ore, and experiment till he strikes that combination of correct proportion of the ingredients of the charge, degree of heat, and length of time in furnace, which will give him the best results.

Let him remember that the principles of fluxing are true for either process.

Preparation of the charge. After having obtained a knowledge of the ore by tests based upon the study of all the preceding, the proper charge is made up. Further on I have given special charges and directions for certain ores, but here I make it only general.

The most convenient and commonly used amount of ore is either one of the standard weights I have so frequently spoken of, viz.: I A. T., 30 grammes, or 480 grains. If a very low grade gold ore is under examination, the above quantities can be doubled, trebled, or quadrupled, or, if necessary, ten or even twenty times the standard weight may be used. But such large quantities, with the fluxes accompanying, are rather difficult to handle, so that it is best to run several charges of the standard weight (or, perhaps, double it), and to unite the resultant buttons by scorification. If it should happen that there is not enough of the ore to make a fair charge (say not half the standard weight), then do not trouble about the crucible process at all, but treat the sample by scorification.

Bi-carbonate of soda, or, as it is commonly called, soda, is used in *every* crucible assay. Although not absolutely necessary, still a mixture of it with carbonate of potash gives somewhat more fluid slags. About one part of the latter to four parts of the former is a good proportion. Considering the ore as one part, the amount of soda (or of soda and carbonate of potash together) varies from one-half to three parts. A safe quantity to use is two parts.

Litharge is also used in all crucible assays save those of ores very rich in lead. The proportion varies from one part to eight parts—one to two parts being the usual range.

Silica is usually employed only for ores full of lime, magnesia, baryta, etc.; in short, for those whose gangues are basic, or when the ore contains no gangue; but I make it a universal rule to use it in *all* crucible assays. For those ores which have no gangues or are basic, it is certainly needed. For those which contain silica I still add it, to be certain to convert the excess of lead over and above that required for the standard weight of lead button into silicate of lead, a most efficient flux. This for ores worked by the litharge process. In those treated by Aaron's process, the added silica converts the soda into silicate of soda (or soluble glass), a good flux, although not so powerful as the lead glass. For pure carbonate of lime and similar gangues, use the same weight of silica as of ore. For pure quartzose ores, one-half the weight of ore. For less pure quartz, three-fourths the weight of ore. In brief, my final advice is, not to be afraid of using plenty of silica in crucible charges.

Borax is best used in the form of borax glass. For quartzose ores, none is absolutely needed, but about 10 per cent the weight of ore does no harm, and seems to help the fusion. Strongly basic ores need 50 per cent of borax glass. As they diminish in lime, etc., down to quartz ores, diminish the borax glass down to 10 per cent.

Charcoal and other reducing agents, and the oxidizing agent, nitre, are to be used according to their several powers and as the ores vary.

It should also be remembered that where gas furnaces are constantly used, there will not be required so much reducing agent in the crucible charges as in those made up for solid fuel furnaces, the flame and heat of the gas furnaces being more reducing.

A cover of salt is to be invariably used.

Select the crucibles, which should be free from cracks and flaws. A good size for the charges to go with the standard weights of ore is $4\frac{1}{8}$ inches wide by $4\frac{3}{4}$ inches deep, outside measurement (size "S" of Battersea make). Clean the crucibles inside if they need it, and number or letter each by means of liquid ruddle, in several places and in large characters, that there may be no difficulty in identifying them after fusion.

Weigh out next the soda for the charge, and brush on to a clean piece of black glazed paper. Next weigh the litharge very carefully, and brush on top of the soda. The silica, weighed approximately, is followed by the ore, charcoal, or other reducing agent, or nitre, each weighed carefully. If sulphur is used, it can come next. Finally, weigh the carbonate of potash, rapidly and but approximately, as it quickly absorbs moisture from the atmosphere; transfer to the paper, and mix everything thoroughly.

Brush the charge into the proper crucible, which it will probably fill two-thirds, tap gently till the contents are level, drop the weighed borax glass on top, and cover with about one-half inch common salt. If nails are used, insert them in the charge before the salt is added.

Another method of charging or "dressing" the crucibles is to pour the soda into the crucible, then the ore, following with the other fluxes, and mixing all together in the crucible with any convenient utensil, as a spatula, spoon, or glass rod.

The marking of the crucibles may be dispensed with if the fusions are made in regular order; but in this case, great care must be exercised, for doubt once entertained as to the identity of any crucible may dispel faith in those following or preceding it.

Running the Crucibles in the Fire.

Have the fire quite hot, and place on the coals the crucible, holding it with the tongs with one hand, and banking it around with the fuel till it sets firmly and uprightly in place. Place the cover on, and surround with coke, charcoal, or coal, as the case may be. Coke works the best when of about the size of an egg, charcoal may be in somewhat smaller pieces. The finer particles of either are often useful when the fire gets too hot, by choking the draft. Arrange the draft and damper so as to permit of a gradually increasing heat.

Occasionally examine the crucible to see if all is working smoothly. In from twenty-five to forty-five minutes the contents of the crucible should be in quiet fusion. The length of time that a charge requires to be thoroughly melted in depends so much upon the ever varying conditions of temperature of the furnace, character of ore, size and nature of

charge, etc., that no exact rule can be given —
only see that the charge *is* fused.

When satisfied of this, remove cover, lift
out the glowing crucible by means of the long-
handled crucible tongs, tap gently, cover, and
let cool, or pour into the scorification mould,
which will hold the button and some of the
slag ; the excess of the latter can run to
waste.

If nails have been used, before tapping or
pouring rinse each nail *in* the slag, tap and
remove.   No lead should adhere to them.

Crucibles that have had their contents
poured out can be employed a second time, or
even more often.   In case of ores that have
shown little or no gold or silver this may do,
but with ores of any richness it is a dangerous experiment.   Accuracy should never be
sacrificed to a spirit of false economy.

Never try to cool a crucible by dipping it
into, or holding it under, cold water, as the but
partially cooled lead is liable to separate into

globules of various sizes, incurring danger of loss.

When stone-cold, break the crucible by striking it a few sharp blows down one side with a hammer. If skilfully done, the crucible will separate in halves the entire length, exposing the lead button. Such a section would appear about as here shown, showing at the bottom the lead button containing the gold and silver, above it the slag, and topping all a layer of fused salt. (The thin body just above the button shows

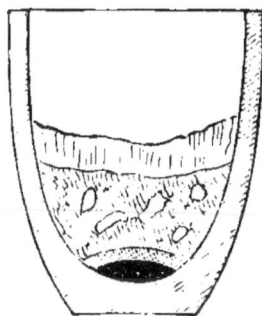

FIG. 129.

how a "matte" would appear were any formed by the fusion of a sulphuret.)

From the broken crucible or the poured charge in the mould save a piece of the slag for future examination and comparison. It should be uniform in color and composition —the former will vary with the character of the ore and proportions of the ingredients of the charge. If the latter has been poured, the crucible should be but little corroded, and smoothly lined with a thin glaze of the slag, and retain no lumps of semi-fused nature or pellicles of lead.

Free the button from the slag, which it should easily and cleanly leave, hammer into shape, as usual, and mark. It may be well to weigh it, so that the assayer may know whether he has used the proper quantity of reducing or oxidizing agent, etc., or to modify the treatment on the same or similar ores in the future.

If the button is too large, it may be reduced by scorification ; but this introduces an additional step, and may not give quite so accurate a result.

Finally, cupel the button, weigh, part, inquart, etc., as previously directed.

With all ores poor in silver, deduct the silver known to be in the litharge, according to the amount of the latter employed.

General Charges.—From what has been previously written of the varying characters of ores the student can see how difficult, if not impossible, it is to give a charge of fluxes, etc., which shall satisfy every ore. Still, there are a few general formulæ useful to have and quite easy to remember, which can be more or less modified to suit any particular ore.

MITCHELL'S CHARGE FOR ALL GOLD AND SILVER ORES.

Ore .... 1 part.
Soda 1   "
Litharge....................... 5 parts.
Borax glass .... 1 part.
Nitre or charcoal and always a cover of salt.

The above in each of the three systems of weights :

	A. T. Weights.	*Gramme Weights.*	*Grain Weights.*	
Ore	1 A.T.	30 grammes	480 grains	=1 oz.
Soda	1 "	30 "	480 "	=1 "
Litharge	5 "	150 "	2,400 "	=5 "
Borax glass..	1 "	30 "	480 "	=1 "
Salt cover.				

A similar but more commonly used proportion is :

```
Ore ........................... 1   part.
Soda .......................... 1    "
Litharge ....... ............. 1⅔   "
```
Borax or silica, nitre or charcoal, and salt **cover.**

Or, to put it more conveniently :

	A.T. Weights.	Gramme Weights.	Grain Weights.
Ore 1	A.T.	30 grammes	480 grains=1 oz
Soda 1	"	30 "	480 " =1 "
Litharge ..1⅔*	"	50 "	800 " =1⅔ "
Salt cover.			

Mr. George L. Stone has published the following as a universal flux for basic silver ores (*i.e.*, those in which the gangue is lime, baryta, etc. — for instance, the three spars, calcspar, heavy-spar, and fluor-spar) :

```
Bi-carbonate of soda........... 9 parts.
Borax glass................... 3    "
Argol ....................... 1 part.
```

" Mix thoroughly, and keep on hand ready for use. For one-third assay ton of ore fill

---

* 1⅔ A.T. or 1.66⅔ A.T. cannot be exactly weighed with the A.T. weights, but the sum of the following weights will approximate it sufficiently : 1 A.T. +0.50+0.10+0.05=1.65.

the crucible about two-thirds full of the flux, adding two or three iron nails if the ore contains much sulphur."

CHAPMAN'S CRUCIBLE FLUX.

"A useful flux, employed largely by the writer during the past ten or twelve years, and which has been found, both in his own practice, and in that of others, to yield good results in all general cases, has the composition given below :

> 3 lbs. carb. soda,
> 2 " dried borax,
> $\frac{1}{4}$ lb. cream of tartar,
> 2 oz. white sugar.

" The re-agents in these proportions must be intimately intermixed. The above quantities will dress from 18 to 20 crucibles, when about 25 grammes of ore are taken for assay." (Chapman's Assay Notes, 1881, p. 31.)

AARON'S GENERAL FORMULA.

(For ores to be worked by his second system. See page 278.)

Ore 1   part.
Soda 3   parts.
Litharge 1   part.
Borax .... $\frac{1}{2}$ part.
Sulphur .... $\frac{1}{10}$ "
Flour .... $\frac{1}{10}$ "
Iron 3   nails.
Glass.
Salt to cover.

" Melt, and leave in strong fire about twenty minutes after fusion." (Aaron's Assaying, 1884, p. 53.)

The above amplified as usual is as follows:

	A. T. Weights.	*Gramme Weights.*	*Grain Weights.*
Ore	1 A. T.	30 grammes	480 grains= 1 oz.
Soda	3 "	90 "	1,440 " = 3 "
Litharge ..	1 "	30 "	480 " = 1 "
Borax	$\frac{1}{2}$ "	15 "	240 " = $\frac{1}{2}$ "
Sulphur...	$\frac{1}{10}$ "	3 "	48 " = $\frac{1}{10}$ "
Flour.....	$\frac{1}{10}$ "	3 "	48 " = $\frac{1}{10}$ "
Iron	3 nails.		
Glass.			
Salt cover.			

The same excellent authority, Mr. Aaron, gives the following charge for " ordinary ores containing little or no sulphuret, some quartz, clay, lime, iron, oxide, etc. " :

Ore........................... 1 part.
Soda .... 1 "
Litharge .... 2 parts.
Dried borax .... 1 part.
Flour........ ....$\frac{1}{16}$ "
Salt cover.

" Fuse quickly ; keep in furnace five to ten minutes after subsidence." (Aaron, page 49.)

Putting the above in the three systems of weights, and reducing the flour one-half, since the quantity given is based on a standard of $\frac{1}{2}$ A. T. ore, we have the following :

	A. T. Weights.	*Gramme Weights.*	*Grain Weights.*	
Ore........	1 A. T.	30 grammes	480 grains	=1 oz.
Soda.......	1 "	30 "	480 "	=1 "
Litharge ...	2 "	60 "	960 "	=2 "
Dried borax.	1 "	30 "	480 "	=1 "
Flour	1 grm.	1 grm.	16 "	
Salt cover.				

The following and concluding charge of

this section is one I have used many times for ores similar to those above spoken of; that is, not containing much sulphuret, or oxidized metal; "dry ores," in short, or, as Mr. Aaron calls them, "ordinary ores." The nature of the gangue is unimportant, as I have gotten as perfect and as vitreous slags from pure limestones as from pure quartzose rock, by using it.

	A. T. Weights.	Gramme Weights.	Grain Weights.
Ore	1 A. T.	30 grammes	480 grains=1 oz.
Bi-carb. soda	1½ "	45 "	720 " =1½ "
Carb. potash	½ "	15 "	240 " = ½ "
Litharge ...	1½ "	45 "	720 " =1½ "
Silica	1 "	30 "	480 " =1 "
Borax glass.	½ "	15 "	240 " = ½ "
Charcoal ...	$\frac{6}{10}$ grms.	$\frac{6}{10}$ "	9¼ "
Salt cover.			

The charcoal is used on the basis of a reducing power of about 25. Time in fire, about one-half hour.

Special charges and directions.—I shall here give charges for certain well known ores, as examples of the varying modes of treatment

These are iron pyrites and its oxide, the sulphurets of copper and their oxides, and galena. All other ores and combinations can be worked by some one of the general or special methods, or some slight modification of one of them.

I. — IRON PYRITES.

Pyrite, the true iron pyrites, is one of the most widely distributed of minerals, is found in the rocks of every age, and is a common and abundant source of gold. There are many varieties of it, and of another and similar mineral, pyrrhotite, which have received many names, as pyrites, iron pyrites, pyrite, marcasite, mundic, bi-sulphuret or bi-sulphide of iron, sulphuret or sulphide of iron, and pyrrhotite, magnetic pyrites, magnetic iron pyrites, magnetic sulphuret of iron, magneto-pyrites, etc. But since the assay treatment makes no mineralogical distinctions, I shall include under the one simple and well known heading of "iron pyrites" any combination of iron and sulphur, with or without a gangue, and direct accordingly.

Iron pyrites has frequently united or associated with it varying amounts of other metals, as nickel, cobalt, copper, zinc, manganese, arsenic, antimony, etc. If in small quantity, none of these need influence the charge or manner of treatment. If in large quantity, making either a compound mineral, or a combination of various minerals, they may necessitate some variation in the mode of treatment.

Since, as stated, iron pyrites is found in rocks of every geological epoch, we may expect to have every kind of gangue ; but, by remembering the rules already given, that acid and basic gangues are to have basic and acid fluxes, no trouble need be feared. Silver, if present, will come down with the gold.

METHOD A.

Desulphurization by a preliminary roasting.

For the ore without any gangue, or for

"concentrates," mix 1 part (1 A. T., 30 grammes, 480 grains) with 1 part of silica.

Mix the above, or the ore alone, if it has a gangue, with its bulk of fine charcoal, and roast according to pages 273–278.

After roasting, add the other constituents, so that the total charge will be as follows :

CHARGE.

	A. T. Weights.	Gramme Weights.	Grain Weights.
Ore (roasted) .	1 A.T.	30 grammes	480 grains = 1 oz.
Bi-carb. soda .	1½ "	45 "	720 " = 1½ "
Carb. potash .	½ "	15 "	240 " = ½ "
Litharge	1½ "	45 "	720 " = 1½ "
Silica	1 "	30 "	480 " = 1 "
Borax glass . .	⅕ "	6 "	96 " = ⅕ "
*Charcoal	¾ grm.	¾ "	11½ "
Salt cover.			

Use an "S" Battersea crucible (4¾ inches deep, 4⅛ inches across, outside measurement). If it is feared that a little sulphuret may remain unoxidized during the roasting, one nail may be pushed into the charge, and is to be removed immediately after fusion.

* Assumed to have a reducing power of 25.

Time in fire about thirty minutes, or, at all events, let remain heating for ten minutes after the charge has settled into quiet fusion.

It is well to bear in mind that iron pyrites (pyrite) loses exactly one-third of its weight by being roasted into the red oxide, so that one part of the sulphuret becomes but two-thirds of a part of oxide. (The varieties of pyrrhotite lose from one-seventh to one-eighth in weight.) Further, that one part of the red oxide keeps oxidized about $1\frac{1}{3}$ parts of litharge, or, more exactly speaking, it oxidizes the charcoal which would reduce $1\frac{1}{3}$ parts litharge.

The charge given, as stated, is for an ore almost, if not entirely, iron pyrites. From such a large percentage it may run down to a very small amount of the mineral, the gangue, of a necessity, increasing accordingly. If the gangue is silicious, diminish the added silica in the charge as the proportion of gangue increases, but do not altogether omit it for even the most silicious ore — retain from $\frac{1}{4}$ to

½ a part. If the ore is limey or otherwise basic, retain the full part of silica, and increase the borax glass to ½ a part.

The charcoal (or any other reducing agent) must, of course, be decreased as the iron pyrites decreases, for the less there is of the latter the less oxidizing power there is in the roasted ore. The actual quantity of charcoal to use on any particular ore is determined by either *guessing* the percentage of sulphuret in the unroasted ore, or by making a preliminary assay to determine the oxidizing power of the roasted ore. Instead of doing the latter additional work, the assayer can come closely enough to the size of the button he wants (15 grammes or 240 grains) by adding charcoal as the *redness of the roasted sample increases*, and *vice versa*, as follows (the charcoal considered to be of a reducing power of 25):

Color of Roasted Ore.	Grammes of Charcoal to Use.
Very deep red	0.850
Deep red	0.800
Medium deep red	0.750
Red	0.700
Medium light red	0.675
Light red	0.650
Very light red	0.625
Very light rose or pink	0.600

The slags from crucible runnings of the ore we are discussing, if the fusions have been properly conducted, will always be homogeneous and glassy in texture, and will be translucent or transparent when not much iron oxide is present, opaque when the oxide preponderates; in color, ranging from very light green to a black, with a greenish or grayish tint accordingly as the iron oxide increases.

METHOD B.

Desulphurization during the fusion.

With ores treated by this method we do not get rid of the sulphur of the iron pyrites by a preliminary operation, but do the desul-

phurizing during the fusion by the oxidation
of all the sulphur (save a sufficient quantity
to bring down a 15 gramme or 240 grain button) by means of nitre.

CHARGE.

For pure pyrites (*i.e.*, free from gangue) and "concentrates."

	A. T. *Weights.*	Gramme *Weights.*	Grain *Weights.*	
Ore 1	A.T.	30 grammes	480 grains	=1 oz.
Bi-carb soda .1½	"	45 "	720 "	=1½ "
Carb. potash. ½	"	15 "	240 "	= ½ "
Litharge 1½	"	45 "	720 "	=1½ "
Silica 1	"	30 "	480 "	=1 "
Borax glass.. ⅕	"	6 "	96 "	= ⅕ "
Nitre........2	"	56½ "	870 "	

Salt cover.

Use Battersea "J" crucible (6¾ inches deep,
4⅜ inches across, outside measurement).

Time, about half an hour.

The oxidizing power of nitre varying, the
amount used may have to be altered somewhat according to its strength, as determined
on page 176. The quantity taken was deduced
from the following calculation — for example,
the weight in grammes: 1 gramme iron pyrites
reduces 8.6 grammes lead (according to the

table on page 263); hence, 1 74 grammes py
rites will reduce about 15 grammes lead; 30–
1.74=28.26= grammes of iron pyrites to be
oxidized.   1 gramme iron pyrites requires for
oxidation from 2 to $2\frac{1}{2}$ grammes nitre; hence,
28.26×2=56.5= grammes nitre to use.

Run in good fire, twenty minutes to fusion,
ten afterward.   Slag brown-black, vitreous,
opaque.   If the button is at all brittle from
presence of sulphur, add lead and scorify.

For other ores, as the iron pyrites diminishes
(and the gangue increases), lessen the silica
and nitre.   The quantity of the latter to use
is determined by experience and experiment,
or by ascertaining the reducing power of the
ore, for which latter see page 256.

METHOD C.

Desulphurization during the fusion.

This method of treating the ore is to use
an excess of nitre, *i.e.*, more than enough to
oxidize all the sulphur, bringing everything
into quiet fusion, and then throwing down

the proper weight of lead by adding a known amount of some reducing agent, as charcoal, or a mixture of charcoal and litharge, or galena and litharge. The reduced lead in its passage down through the molten charge absorbs the gold and silver.

CHARGE.

For iron pyrites with no gangue, or concentrates.

	A. T. Weights.	Gramme Weights.	Grain Weights.
Ore...........	1 A.T.	30 grammes	480 grains = 1 oz.
Bi-carb. soda.	1½ "	45 "	720 " = 1½ "
Carb. potash.	½ "	15 "	240 " = ½ "
Litharge	6 "	175 "	2,880 " = 6 "
Silica	2 "	60 "	960 " = 2 "
Borax glass..	1 "	30 "	480 " = 1 "
Nitre.........	2½ "	75 "	1,200 " = 2½ "

Salt cover.

Use a "J" crucible.

Begin with a very gentle heat, and gradually bring up to a full red-heat (thirty-five minutes). When in full fusion add the following weight of charcoal wrapped in tissue paper (considering the weight of the paper as charcoal) :

1 gramme, or 15½ grains.

When again in·quiet fusion (five minutes), remove and tap, or pour.   Slags brownish-black, vitreous, opaque.

For less pyritic ores use less nitre.

METHOD D.

Desulphurization with iron.

CHARGE.

Make up the ore with fluxes as follows:

	A. T. Weights.	Gramme Weights.	Grain Weights.
Ore	1 A.T.	30 grammes	480 grains = 1 oz.
Bi-carb. soda.	1½ "	45 "	720 " = 1½ "
Carb. potash.	½ "	15 "	240 " = ½ "
Litharge	1½ "	45 "	720 " = 1½ "
Silica	1 "	30 "	480 " = 1 "
Borax glass..	⅕ "	6 "	96 "
Iron nails, 6.			
Salt cover.			

Use "S" crucible.   Time, one-half hour.

Tie the nails with wire together, and stick the bunch into the charge, points down. After fusion, should any "matte" be formed, it must be scorified, with the lead button, to a pure and malleable condition.

Slags brown-black, vitreous, opaque.

METHOD E.

Converting the iron pyrites into matte.
Aaron's process (modified).

CHARGE.

For concentrated pyrites.

	A. T. Weights.	Gramme Weights.	Grain Weights.	
Ore	1 A.T.	30 grammes	480 grains=	1oz.
Bi-carb. soda .	3 "	90 "	1,440 "	=3 "
Litharge	½ "	15 "	240 "	= ½ "
Silica	1 "	30 "	480 "	=1 "
Borax glass..	½ "	15 "	240 "	= ½ "
Nitre	⅖ "	12 "	18½ "	

Iron nails, 6.

Salt cover.

Use " S " crucible. Time forty minutes.

The nitre is to oxidize the excess of sulphur; *i.e.*, that over and above the amount necessary to reduce all of the litharge, and yet not enough of this re-agent is used to prevent the formation of a matte.

Slag coal-black, vitreous, opaque.

Mr. A. H. Low, of Argo, Colo., has given me a good hint in the crucible running of sulphuret ores, which I herewith note. Make the fusion in the usual manner, and when it is supposed to be completed, take out, and pour off as much of the slag as possible without losing any of the lead. The button can now be easily seen, and if all the sulphur has not been driven off, replace the crucible in the fire *at an angle*, and scorify, as it were, till the sulphur has gone ; take out, pour, and the result will be a clean button.

For ores which are much richer in gold than silver, and in which it is not desired to determine the latter, the operation of inquartation of the resultant bead can be dispensed with by putting in the charge before running a piece of pure silver of the proper weight ; it will help to collect the gold, moreover.

Comments on the five preceding processes.— With these, as with others, each assayer will find some certain one will, with him, work better, and give higher results than the others. He can then practice until he ascertains which one suits him the best, and always employ that.

My own experiments, and those of my students, tend to prove that each of the first three methods is pretty sure to bring down all the gold ; that the fifth method must be worked very carefully to obtain correct results, and that the fourth is the most unsatisfactory of all. Further, that the first method gives higher silver than any of the others, which supports the theory that, with sulphur in an ore during its fusion, oxy-sulphurets are formed, which drag silver with them into the slag.

Many ignorant assayers insist that from *unroasted* sulphuret of iron ores, no gold will be obtained by crucible fusion. This extraordinary idea, which, however, seems to spread, is entirely unfounded, or, at best, is based upon botchy experiments. The student may rest assured that he can, with careful working, extract all the gold from an unroasted ore by any one of the nitre methods.

II. — OXIDE OF IRON.

One of the results of the decomposition of
iron pyrites is oxide of iron, existing either as
limonite, which is the oxide with water, or as
hematite, the oxide without water.   It is even
more widely distributed than iron pyrites, and
since the latter is one of the most common
sources of gold, so likewise is oxide of iron,
with the advantage that that precious metal
is retained therein in a free-milling condition.
It is so universal that any brownish, yellowish, or reddish coating on an ore is almost
certain to be partially, if not entirely, oxide
of iron.

The ore can be treated in two ways, by the
charge given for the iron pyrites after roasting (on page 299), remembering that there
is a full part of this ore, against three-fourths
of a part of the roasted sulphuret, or by
Aaron's process.

CHARGE.—LITHARGE PROCESS.

	A. T. Weights.	Gramme Weights.	Grain Weights.
Ore	1 A. T.	30 grammes	480 grains = 1 oz.
Bi-carb. soda	1½ "	45 "	720 " = 1½ "
Carb. potash	½ "	15 "	240 " = ½ "
Litharge ...	1½ "	45 "	720 " = 1½ "
Silica	1 "	30 "	480 " = 1 "
Borax glass.	⅛ "	6 "	96 " = ⅛ "
Charcoal ...	1 grm.	1 "	15½ "
Salt cover.			

The above charge is, for the oxide ore, quite free from gangue. As has before been stated, diminish the silica and charcoal as the gangue increases.

It may be as well to add a cautionary note, to the effect that limonite, the hydrous oxide (yellow or brown in color), contains about 14½ per cent. of water of combination, while hematite, the red oxide, contains none whatever; hence use more charcoal, or other reducing agent, for the latter than for the former.

CHARGE.— AARON'S PROCESS (MODIFIED).

	A. T. Weights.	Gramme Weights.	Grain Weights.
Ore	1 A. T.	30 grammes	480 grains = 1 oz.
Bi-carb. soda	3 "	90 "	1,440 " = 3 "
Litharge	½ "	15 "	240 " = ½ "
Silica	1 "	30 "	480 " = 1 "
Borax glass.	½ "	15 "	240 " = ½ "
Sulphur	½ "	15 "	240 " = ½ "
Flour	4/10 "	12 "	18½ "

Iron nails, 6.

Salt cover.

Many times the assayer will meet ores which are mixtures of iron pyrites, and its decomposed substitute, iron oxide. Such must be treated by the methods given for the former mineral. If the pyrites is in very small quantity, Process A, *without* a preliminary roasting, may be used (in this case employ but from one-third to one-half the amount of charcoal), as the fluxes will do all the necessary desulphurizing.

III.— SULPHURETS OF COPPER.

Including sulphuret of copper (chalcocite, copper glance, vitreous copper, copper sulphide), which when pure contains 79.8 per cent of copper and 20.2 per cent sulphur, and which is more liable to carry silver than gold; copper pyrites (chalcopyrite, sulphuret of copper and iron), a valued source of the precious metals, and which is composed of about equal parts of copper, iron, and sulphur; bornite (purple copper ore, variegated copper ore, variegated copper pyrites, erubescite, "horse-flesh" ore, sulphuret of copper and iron), an ore similar to the preceding, but with less iron; gray copper ore ("fahlerz," tetrahedrite), a sulphide of copper and antimony with smaller and varying amounts of other sulphurets; and finally, all other and rarer sulphurets with copper as an important ingredient, such as Barnhardite, Bournonite, Carrollite, Covellite, Harrisite, Stromeyerite, and Tennantite.

These ores can be treated by any one of the five methods given for the sulphuret of iron, and which I repeat here, with some alterations in the proportions of the ingredients of the various charges, and, for the sake of space, in tabular form.

Even these charges will have to be modified more or less to suit any particular sample, for it may be a mixture of two or more of the above-named ores, but they will serve as representative methods. Salt cover to all, as usual.

	METHOD A.	METHOD B.	METHOD C.	METHOD D.	METHOD E.
Ore.	1 A. T. 30 grms. 480 grains. Roasted.	1 A. T. 30 grms. 480 grains.	1 A. T. 30 grms. 480 grains.	1 A. T. 30 grms. 480 grains.	1 A. T. 30 grms. 480 grains.
Bi-carb. Soda.	1½ A. T. 45 grms. 720 grains.	1½ A. T. 45 grms. 720 grains.	1½ A. T. 45 grms. 720 grains.	1½ A. T. 45 grms. 720 grains.	3 A. T. 90 grms. 1,440 grains.
Carb. Potash.	½ A. T. 15 grms. 240 grains.	½ A. T. 15 grms. 240 grains.	½ A. T. 15 grms. 240 grains.	½ A. T. 15 grms. 240 grains.	o
Litharge.	2 A. T. 60 grms. 960 grains.	2 A. T. 60 grms. 960 grains.	6 A. T. 175 grms. 2,880 grains.	2 A. T. 60 grms. 960 grains.	½ A. T. 15 grms. 240 grains.
Silica.	1 A. T. 30 grms. 480 grains.	1 A. T. 30 grms. 480 grains.	2 A. T. 60 grms. 960 grains.	1 A. T. 30 grms. 480 grains.	1 A. T. 30 grms. 480 grains.
Borax glass.	⅕ A. T. 6 grms. 96 grains.	⅕ A. T. 6 grms 96 grains.	1 A. T. 30 grms. 480 grains.	⅕ A. T. 6 grms. 96 grains.	½ A. T. 15 grms. 240 grains.
Charcoal.	8/10 grm. 12⅓ grains.	o	After fusion 1 grm. 15½ grains.	o	o
Nitre.	o	1¾ A. T. 50 grms 770 grains.	2¼ A. T 67 grms. 1,080 grains.	o	3/10 A. T. 9 grms. 114 grains.
Nails.	o	o	o	6	6

Copper pyrites — that is, the sulphurets of iron and copper partially oxidized — should be treated like the unchanged pyrites. When entirely converted into the oxides of copper and iron use Method A of the sulphuret of copper charge given above, and Aaron's charge on page 312.

IV.—OXIDES OF COPPER.

The oxides of copper are the red oxide (cuprite), with 88.8 per cent copper, and the black (melaconite, tenorite), with 79.8 per cent of that metal. As oxidized minerals there are the hydrated oxy-carbonates, malachite (green carbonate), and azurite (blue carbonate), with 57.4 and 55.3 per cent of copper respectively.

The two oxides differ chiefly in this, that the black oxide is oxidizing, *i.e.*, giving up its oxygen during fusion, while the other does not do so. Hence, extra reducing agent must be added to the former.

Use either the litharge or Aaron's process.

For the former, make up the charge given as Method A for sulphurets of copper, using $\frac{8}{10}$ grm. ($12\frac{1}{3}$ grains) charcoal for the black oxide, and $\frac{1}{2}$ grm. ($7\frac{3}{4}$ grains) for the red.

For Aaron's process, use the charge on page 312.

Malachite, azurite, and other oxidized ores containing not as much copper as the above two oxides, will need but the regular amount of litharge for the litharge process ($1\frac{1}{2}$ A. T., 45 grammes, 720 grains).

Chrysocolla, the silicate of copper, calls for no especial remark, save only that the charge will not require quite so much silica.

V.— GALENA.

Galena (galenite, "pyrites of lead," sulphuret or sulphide of lead) is so frequently a source of silver, or, at least, so many times associated with other minerals that do carry silver, that its assay is often called for. Sometimes, although not often, galena is rich in gold and quite free from silver; but whether

auriferous or argentiferous, it makes no difference in the manner of treatment.

The silver in this ore varies from a mere
trace to as high as $1,500 per ton. It is
claimed that when galena contains over 0.1
per cent silver, that it is due to the presence
of a true silver mineral. This is probably so;
at all events, one can often pick out from the
cleavage of a rich specimen particles of argentite or horn-silver.

For the assay treatment, use Method B,
p. 289, with the litharge $\frac{1}{2}$ part (meaning by a
part 1 A. T., 30 grammes or 1 oz) and the
nitre $\frac{6}{10}$ of a part. Or Method C, with 1 part
litharge and $\frac{3}{4}$ part nitre. Or, finally, Method
D, with $\frac{1}{2}$ part litharge.

An approximate assay for silver in galena
can be made by cupelling the buttons obtained
in the *lead assay*, etc., and making the proper
calculations, remembering that grammes and
grains, and not A. T. weights, are used in
such assays.

CHAPTER II.

COPPER ORES.

OCCURRENCE.—Copper is found both native and in combination with many elements, principally with sulphur as a sulphide or sulphuret, with oxygen as an oxide, and with carbon, hydrogen and oxygen as a hydrated carbonate. It has also been discovered associated with most of the metals, common or rare.

It is obtained for the arts and manufactures mostly from the following ores :

1. Native copper (copper, sometimes accompanied by silver), when pure, 100 per cent.

2. Cuprite (red oxide of copper), with 88.8 per cent copper.

3. Melaconite (black oxide of copper), with 79.8 per cent copper.

4. Azurite (blue carbonate of copper), with 55.2 per cent copper.
5. Malachite (green carbonate of copper), with 57.4 per cent copper.
6. Chalcocite (sulphide of copper), with 79.8 per cent copper.
7. Chalcopyrite (sulphide of copper and iron), with 34.6 per cent copper.
8. Tetrahedrite (gray copper ore), copper variable, normally contains about 38 per cent.*

Assay.—Of the many *dry* methods for the testing of copper ores, it may safely be said that no single one is very accurate. The various metallurgical works usually have processes or modifications of processes peculiar to themselves, but which are always more or less imperfect. Many of these processes are complicated, and require great skill with constant practice.

I have then thought it best to specify but

---

* See appendix for more extended list of copper minerals.

three assay methods, they being representative ones.

I. METHOD FOR NATIVE COPPER.

(As a simple mixture of rock and metallic
copper.)

Here the only action is fusion.

CHARGE.

Ore..................... 10 grammes or 160 grains.
Bi-carbonate of soda 20   "   " 320   "
Carbonate of potash 5   "   "   80   "
Borax glass............. 1   "   "   16   "
Salt and charcoal cover.

Sample ore as usual. Mix charge and pour
in "U" crucible. Put cover of $\frac{1}{2}$ inch salt
and then 1 inch of wood charcoal. Cover and
heat intensely for twenty to thirty minutes.
After cooling, break crucible and clean button from slag. Divide the weight of button
obtained by 10 or 160, and multiply this result
by 100 for percentage of copper,

II. METHOD FOR OXIDES AND CARBONATES OF
COPPER, FREE FROM SULPHUR.

Here the action is reducing, followed by
the collection of the copper globules into one
button.

CHARGE.

Ore .... 10 grammes or 160 grains.
Black flux substitute 30   "   " 480 "
Borax glass............. 5   "   " 80 "
Argol 2   "   " 32 "
Salt and charcoal cover.

Use a " U " crucible, chalk-lined. Cover,
heat gradually for twenty minutes, then
increase to white-heat for forty minutes. Remove, tap, and let cool. Results approximate,
the error augmenting by the presence of other
metals.

III. METHOD FOR SULPHIDES OF COPPER, WITH
ARSENIC, ANTIMONY, LEAD, MERCURY, ZINC,
ETC.

The first step, concentration, is to bring
down into a matte all the copper, and to get
rid of obnoxious lime or baryta gangues,

Ore, according to

    richness........ 10 to 30 grms., or 160 to 480 grains.

*Iron pyrites .... 2 " 6 " " 32 " 96 "

Borax glass ..... 8 " 24 " " 128 " 384 "

Salt cover.

Mix as usual and transfer to crucible. If lead is present, put in a couple of nails. Fuse in hot fire, remove when finished, take out nails if any have been used, cool, break away the slag from the matte, and treat the latter as follows :

The second step, roasting, is to expel the sulphur and the volatile metals, arsenic, antimony, zinc, or mercury, converting the copper into an oxide. Great care must be employed here, hence observe the directions given under *Roasting* in the crucible process for gold and silver ores, pages 273 to 278, using coke in place of charcoal, and *no silica*.

The third step is reduction, or bringing the copper from its state of oxide to that of a metal.

---

* Containing no copper; test with nitric acid and ammonia. If pyrites is in the ore to any extent, no extra amount need be added.

CHARGE.

Ore (roasted)............30 grammes or   480 grains.
Black flux substitute.....90   "   " 1,440   "
Borax glass 15   "   "   240   "
Lime glass .... 7½   "   "   120   "
Red oxide iron 3   "   "   48   "
Salt cover.

(If 10 grammes or 160 grains ore is used, reduce the above quantities two-thirds.)

Mix the ore, oxide iron, and one-third the black flux substitute, transfer to the crucible, settle down, then add the remaining two-thirds of the black flux substitute, the borax glass, lime glass, and common salt, in consecutive layers, and on top of all a piece of coal about the size of a hazel nut. Heat slowly at first, then intensely, remove, cool, detach button, etc. If it seems red and pure and is malleable, weigh and calculate percentage. If not, then there comes the fourth step, purification or refining.

Two large cupels are well heated in the muffle; into each a piece of pure lead 3 grammes or 48 grains is placed, and the muffle is closed. When the leads have melt-ed, open the door, and into one cupel drop the impure button, into the other a piece of pure copper of the same weight. Let them remain till "brightening" occurs, indicated by a peculiar green color. As soon as this has happened, cover cupels with coke or coal dust, take out and cool in water. The loss the pure copper has sustained in cupellation is supposed to be the same as the button loses, and is to be added to that of the latter. Weigh and calculate on first weight of ore taken. Results moderately accurate.

Read Mitchell, and Bodeman and Kerl on copper assays.*

* Several criticisms have been made upon this chapter to the effect that it is altogether too brief. I have made no endeavor to extend it in this revised edition, for the reasons (one of which has been previously stated) that the fire processes are *not accurate*, and that they are gradually being substituted everywhere by the more exact volumetric and electrolytic methods which I describe in full detail in the appendix. The dry assay methods described are as good as any, and cover the three classes of copper ores common to this country, *i.e.*, free metal, oxidized, and sulphuretted ores.

CHAPTER III.

LEAD ORES.

OCCURRENCE.—Lead is very rarely found
native (that is, as the pure metal), but occurs
combined with various elements, as antimony,
arsenic, carbon, chlorine, chromium, molyb-
denum, oxygen, phosphorus, selenium, sulphur, tellurium, tungsten, vanadium, etc.
Combinations of some of the above elements .
with each other and with lead exist, either
alone or associated with such metals as cobalt,
copper, gold, iron, mercury, nickel, silver,
zinc, etc.

Many of these compounds are merely mineralogical curiosities, and will not be considered here.

The *important workable* ores are the following :

1. Galenite (galena, sulphide or sulphuret of
   lead), when *pure* consisting of 86.61 per
   cent lead and 13.39 per cent sulphur.

2. Cerussite (white lead ore, carbonate of lead), containing 77.52 per cent lead.

3. Minium (red oxide), with 90.80 per cent lead.*

ASSAY.—The method of assaying a lead ore depends upon the nature of the ore.

I. METHODS FOR GALENA.

(Also for selenides, sulphates, and for galena containing antimony and arsenic.)

A. By Crucible Fusion in Furnace.

1. With bi-carbonate of soda and metallic iron.

CHARGE.

Ore .... 10 grammes, or 160 grains.
Bi-carbonate of soda..... 25   "   " 400   "
Carbonate of potash..... 10   "   " 160   "
3 iron nails or
3 loops of iron wire.
Salt cover.

Prepare the sample according to the directions given on pp. 185–191.

---

* For more complete list of lead minerals see appendix.

Weigh first the carbonates, then the finely pulverized ore, and mix thoroughly on glazed paper.

(Read the notes on the "Crucible Assay of Gold and Silver Ores," pp. 257–258.)

Brush into a lettered or numbered small sand crucible (size " U " of Battersea make), and settle contents down.

If there is considerable pyrites in the ore, sprinkle now over the surface of the charge one gramme of finely powdered borax glass.

The three iron nails (eight-penny) are to be held together by their heads with iron wire (No. 16), and then inserted, points down, in the crucible, leaving a loop of the wire hanging over the edge that the nails may be easily and quickly withdrawn when the operation is concluded. If wire only is used, bend a piece of the No. 16, about six inches in length, in the form of a horse-shoe with a loop above, and in the loop hang two smaller pieces bent in the form of hair-pins ; let all six points be about on a level. Insert into the charge.

Finally pack on the surface of the charge and around the nails or wire one-half inch of dry salt.

Place the crucible in a moderately hot fire, *cover* and surround with coke.

This process will require twelve to fourteen minutes.

When fusion is complete, take off the cover, remove crucible from fire, then by means of small tongs stir the nails or wire loops around in the molten mass once or twice, and *while in the hot fluid* tap them against the side of the crucible, then withdraw them, tap gently the crucible and cover. All this should be done as rapidly as possible.

When cold, break and hammer lead into shape as usual.

The weight of the button, multiplied by ten (or divided by sixteen and multiplied by ten), gives the ore's percentage of metallic lead.

Tests of Good Work.—After fusion the interior of the crucible should be smooth and have no half-fused portions adhering to the sides. The charge should be well settled to the bottom and have an even surface. The slag should be uniform in character, and of a purplish-black color. The lead should be at the bottom in *one* button, and be perfectly malleable. A glistening button indicates undecomposed galena ; a brittle one the presence of antimony, arsenic or iron.

The alkaline carbonates act mainly as fluxes, but a portion of the lead they convert into a double sulphide of lead and soda (or potash), which the iron desulphurizes, forming sulphide of iron and metallic lead.

In order to learn the proper running of this lead assay, it will be well for the student to perform it at least *ten* times on the same ore.

2. With black flux substitute and metallic iron.

CHARGE.

Ore...................... 10 grammes or 160 grains.
Black flux substitute 35   "   " 560   "
3 iron nails or
3 loops of iron wire.
Salt cover.

Treat in same manner as for the first method. Let remain in fire twelve or thirteen minutes.

Add one gramme or fifteen and one-half grains of borax glass to pyritic ores.

The carbon of the flour of the black flux substitute exerts an additional reducing action.

Perform this assay a number of times for practice.

The remarks given under the first method are applicable here.

3. With cyanide of potash.

CHARGE.

Ore_____ 10 grammes or 160 grains.
Cyanide of potash_____ 30 " " 480 "
Salt cover.

Use a "D" Battersea crucible. Ram into the bottom of the crucible 10 grammes cyanide, above this pour on the charge, cover first with 5 grammes cyanide, lastly the salt. Time, 14-15 minutes, low red heat, 5 minutes at slightly higher temperature.

Half-a-dozen runnings will be sufficient for this method.

These three methods can be performed with satisfaction in Fletcher's gas furnace (p. 74).

B. By Crucible Fusion in Muffle.
4. With bi-carbonate of soda and argol.

CHARGE.

Ore........................	10 grammes	or	160	grains.
Bi-carbonate of soda	15 "	"	240	"
Carbonate of potash	10 "	"	160	"
Argol.....................	7 "	"	112	"
Flour.....................	5 "	"	80	"
Borax glass	3 "	"	48	"

2 loops of iron wire.
Salt cover.

Mix the ore, soda, argol and flour, and pour into a small sand crucible large enough to stand in the muffle used. Sprinkle over the charge the fine borax glass, insert two pieces of iron wire bent as hair-pins, and tamp down with from $\frac{1}{4}$ to $\frac{1}{2}$ inch dry salt. Use no cover.

Have the muffle at a bright red heat, and place the crucible or crucibles therein; after about ten minutes of good heat, increase the temperature for twenty-five minutes longer, when the contents of the crucibles should be in perfect fusion.

Take out, remove wires, tap, and let stand covered till cool, do not pour, break, hammer button, etc.

The size of the crucibles used will depend upon the height of the interior of the muffle, and the fact that the muffles usually employed are small very often either necessitates a smaller charge, or renders it impossible to use this process.

For this work, one may either use the very small crucibles of the ordinary form, surrounding each crucible with a little cup or platform of fire-clay and sand mixed up with borax water, that it may stand securely in the muffle, or the special form for muffle fusions shown in fig. 73. The latter is recommended.

C. By Fusion in Scorifiers.

This is a modification of the lead assay

designed to be used where the muffles are not large enough to admit of crucibles.

It is simply a substitution of scorifiers for crucibles, using the same charges (reduced one-half in quantity) as for crucible work, and employing the muffle.

Half the third charge works well, thus :

5. With cyanide of potash.

CHARGE.

Ore...................... 5 grammes or 80 grains
Cyanide of potash 15   "   " 240   "
Salt cover.

For a " J " Battersea muffle, employ a 3½ inch Battersea scorifier. Have the muffle red hot, introduce scorifier, cover with 3½ inch circular crucible cover ("G" of Battersea), heat moderately for ten minutes, then intensely for twenty. Remove cover, take out scorifier, do not pour but let cool covered, break, and shape lead button. Multiply weight by twenty (when grammes are used), or divide it by eight and multiply the quotient by ten (when grains are used) for percentage.

Comparison of Processes.—The cyanide of potash process in crucibles gives the highest results, the buttons are clean and malleable, and the slags almost always uniform. I have found it the one most quickly learned, and so, on all accounts, I give it the preference.

The fourth process (crucible in muffle) comes next in percentage of lead obtained.

The second process (black flux substitute) ranges next, and is quite satisfactory to work.

Very close in results to the preceding, is the third process in scorifiers (No. 5, half charge).

The first process (bi-carbonate of soda) gives lower results than any of the others.

A method with ferrocyanide of potash, that is sometimes used, I have omitted entirely, on account of its inaccuracy.

II. METHODS FOR OXIDES AND CARBONATES.

(Cerussite, minium, etc.)

By crucible fusion in or out of muffle.

6. With soda, potash and argol.

CHARGE.

Ore.....................10 grammes or 160 grains.
Bi-carbonate of soda 15   "   " 240   "
Carbonate of potash 5   "   " 80   "
Argol................... 5   "   " 80   "
Salt cover.

Mix as usual, and transfer to small crucible. Cover, if fusion is made in the open fire, but not if the muffle is used. Heat gradually for about fifteen minutes, then somewhat more strongly till fusion ensues. Take out, pour or not, as desired. Cover if left in crucible to cool.

Action reducing—the oxygen of the ore is seized by the carbon of the argol, leaving metallic lead.

7. With soda, argol and borax.

Prepare a flux, in quantity, of the following ingredients :

CHARGE.

   2 parts bi-carbonate of soda.
   2 parts argol.
   1 part common borax, in powder.
   1 part flour.

Have the above well mixed, then sifted, and keep ready for use.

Fill about two-thirds full a so-called "5 gramme" crucible (page 112), with the above flux, add 5 grammes, or 80 grains, of the ore, and mix in crucible. Put in muffle without cover. Keep the heat as low as possible, without letting it get too cold.

If the ore shows sulphurets, put in a nail or two.

If the ore is quite calcic or barytic, make the borax 1½ parts and the flour ¾ of a part.

For ores containing much manganese, add to flux a little more borax and flour.

8. With cyanide of potash.

CHARGE.

Ore_____ 5 grammes or 80 grains.
Cyanide of potash_____ 30 " " 480 "
Salt cover.

Mix as for charge 3 on page 330.

Fuse at low heat for 20 minutes, then end with higher heat for 5 minutes. Make sure that all bubbling ceases,

Concluding Remarks.—At the best, the assay of lead ores is inaccurate, mainly on account of the volatility of the lead itself, though in the case of galenite ores, it is supposed that the galena begins to sublime before the decomposition is effected.

Also, the lead button is liable to contain antimony, iron, and zinc from the ore, or iron from the nails, wire, or iron salts employed in the assay.

Try then to avoid an unnecessarily high heat, remove assays as soon as fusion is obtained, and use covers as much as possible.

The latest investigations on the fire assay of lead tend to demonstrate that cyanide of potash is the *best* reducing flux for *all* the ores of lead, (corroborating the remarks on page 334,) not only for the sulphides, oxides and carbonates, but for the rarer minerals, as the sulphate (anglesite), the phosphate (pyromorphite) and others. A final note of caution is to the effect that the *greater* the amounts of *reducible* metals a lead ore contains (such as

copper, arsenic, antimony, etc.) the greater
the error.   These elements may predominate
to such an extent as to make the fire methods
inapplicable.   For while there may be a com-
pensation of errors (the loss of lead by volatilization counteracted by the addition of the reduced metals), yet too much uncertainty as to
the results would make them practically
worthless.

APPENDIX.

SECTION I.

SPECIAL METHODS.

I. ASSAYING OF THE VARIOUS MINERALS CONTAINED IN AN ORE.

It is sometimes desirable to know *where* the gold and silver are located in an ore, that is, which minerals carry them to the greater extent.   This can not always be done, for the various minerals may be so thoroughly commingled that separation will be impossible. But in other cases, it can be done with success.   For example, an ore is found to be made up of three distinct minerals, blende, galena and pyrites in a quartz gangue.   Weigh the piece selected, and crush roughly in a mortar, taking care not to lose any, and pour out on a clean surface, as a sheet of white paper.   With a pair of pincers pick out such pieces of the quartz as show none of the minerals mentioned, and reject them. Then it will be comparatively easy to put aside, in three piles, the minerals, each quite free from the other two. By carefully crushing the remaining mixed pieces, the entire lump will finally be separated into its three component valuable minerals and the worthless gangue.

Weigh each lot, and assay the whole of each or fractions thereof.

To show method of calculation, I give the following

EXAMPLE.

Weight of sample of ore, 500 grammes,
Which was composed of:

$$\begin{array}{ll}
\text{Pyrites,} & \text{40 grammes.} \\
\text{Blende,} & \text{60 \quad "} \\
\text{Galena,} & \text{100 \quad "} \\
\text{Quartz gangue,} & \text{300 \quad "} \\
\hline
& \text{500 grammes.}
\end{array}$$

$$\begin{array}{llll}
\text{Percentage of} & \text{pyrites} = \frac{40}{500} \times 100 = & 8 \\
\text{"} & \text{"} & \text{blende} = \frac{60}{500} \times 100 = & 12 \\
\text{"} & \text{"} & \text{galena} = \frac{100}{500} \times 100 = & 20 \\
\text{"} & \text{"} & \text{quartz} = \frac{300}{500} \times 100 = & 60 \\
\hline
& & & 100
\end{array}$$

Pyrites.—The 40 grms gave 10 mgrms gold—no silver:
$\frac{10}{600} \times 29.166 = 0.58332 = \frac{58}{100}$ of an ounce per ton.

Blende.—10 grms gave 4 mgrms silver—no gold: 4×6
$= 24$; $\frac{24}{500} \times 29.166 = 1.39$ oz. silver per ton.

Galena.—20 grms gave 160 mgrms silver—trace of gold:
$160 \times 5 = 800$; $\frac{800}{500} \times 29.166 = 46.6$ oz. silver per ton.

II.   ASSAYING OF ORES CONTAINING FREE GOLD OR FREE SILVER.

The average ore does not carry the precious metals in the free state.   But when they are present in such form, proceed as follows:

Crush the sample selected, having first weighed it.   Pulverize as usual, and sift, using the box sieve.   As a result we shall have two things, the finely powdered siftings below and more or less free metal in scales on the sieve.

Weigh the scales, and, as a check, the siftings.   The weight of the latter should be but a trifle less than the difference between the original weight and that of the scales, if care has been taken in pulverizing.

If not too large an amount, *all* the scales

should be wrapped in pure lead-foil, and cupelled directly. If quite an amount be present, simply scorify down in the usual manner.

Take a weighed fraction (one-half, one-tenth, one-twentieth, as the case may be) of the siftings, and assay by scorification or crucible process. The number of milligrammes gold and silver obtained are each to be multiplied by the proper fraction to ascertain the amounts present in the entire bulk of the siftings. Add the gold thus calculated to be in the whole of the siftings to the amount found to be in the scales, and the same with the silver. If the metal is all free gold or all free silver, the calculations are still simpler.

The following example explains itself, and illustrates the methods of calculation. Factors, depending upon the relation of the assay ton of 29.166 grammes to the amount of sample taken, may or may not be used, as desired. If a fixed quantity of ore, as 100 grammes, is always taken, the factors of $29.166 \div 100 =$

0.29166, or of 100 ÷ 29.166 = 3.428, will, of course, always be constant.

Total weight of sample is ....320.000 grammes

Of which the scales weighed .... .260 "

Hence the siftings weighed.... ...... 319.740 "

Factors ....29.166÷320= 0.09114,

Or ....320÷29.166=10.97102.

Scales : the 0.260 grms. produced.. $\begin{cases} 208 & \text{mgrms. silver,} \\ 9.1 & \text{`` gold,} \\ 217.1 & \text{`` total.} \end{cases}$

208÷320=0.65 ;

then 0.65×29.166,
  or 208×0.09114, } =18.95=oz. silver per ton of scales.
  or 208÷10.97102,

9.1÷320=0.0284375 ;

then 0.0284375×29.166,
  or 9.1×0.09114, } =0.83=oz. gold per ton of scales.
  or 9.1÷10.97102,

Siftings : 319.74 grammes,

  30 grammes produced $\begin{cases} 51.75 \text{ mgrms. silver,} \\ 1.50 \quad \text{`` gold,} \\ \overline{52.25} \quad \text{`` total.} \end{cases}$

51.75÷30=1.725 ; 1.725×319.74=551.5515 ; 551.5515÷320= 1.7236 ;

then 1.7236×29.166,
  or 551.5515×0.09114, } =50.2=oz. silver per ton of
  or 551.5515÷10.97102, siftings.

1.50÷30=0.05 ; 0.05×319.74 = 15.987,

15.987÷320 =0 04996 ;

then 0.04996×29.166,

or 15.987×0.09114, } =1.45=oz. gold per ton of siftings.

or 15.987÷10.97102,

Silver: in scales,   18.95 oz., @ \$1.29 per oz.,=\$24.44 per ton.

" siftings, 50.20 " " " " = 64.75 "

" total, 69.15 " " " " =\$89.19 "

Gold: in scales, 0.83 oz., @ \$20.67 per oz.,=\$17.15 per ton.

" siftings, 1.45 " " " " = 29.97 "

" total, 2.28 " " " " =\$47.12 "

Total value of scales ....\$24.44+\$17.15=\$ 41.60 per ton.
Total value of siftings ...\$64.75+\$29.97=\$ 94.72 "
Total value of ore ....\$89.19+\$47.12=\$136.31 "

Check.*

0.260 : 208 :: 29.166 : x, 0.260x=29.166×208=6066.928, x
=6066.928÷0.260=23334.338, 23334.338×1.29=30101.29.
0.260 : 9.1 :: 29.166 : x, 0 260x=29.166×9.1=265.4106, x
=265.4106÷0.260=1020 81, 1020.81×20.67=21100.14.

30101.29+21100.14=51201 43.

0.260 tons scales, @ \$51,201.44 per ton=\$13,312.37
319.740 " siftings, " 94.72 " = 30,285.77

320 tons original ore, worth \$43,598.14 ; or
1 ton is worth \$136 24.

*The idea of the check or verification is taken from C. H. Aaron's excellent little work on assaying, and consists in considering the ore as made up of two kinds of material, assaying each part as though it was the original ore, and from the number of *imaginary tons* of each kind, calculating the actual value of a ton of the mixed or original ore.

III. ANALYSIS OF COPPER ORES.

The *wet* process or analysis of copper ores is so much more accurate than the *dry* process or assay, that it should always be employed when practicable.

As in the assay of copper, so in the analysis, there are many methods, here included under three heads ; volumetry, gravimetry and electrolysis. While there are good ways of determination among the first two classes, yet I prefer one in the third, owing to its simplicity, accuracy, freedom from intricate calculation, and the ease with which it can be acquired.

The process I now describe is known as the *Luckow method*, and consists, briefly defined, in dissolving the copper out of its combinations by means of acids, and then depositing it as the metal itself upon another metal, platinum, by the action of an electric current.

It makes no difference whatever in this method, how the copper is originally combined, whether as a sulphide or in a mixture of sulphides of other metals, an oxide or carbonate, a matte or an alloy; the copper comes out as metallic copper in any case. *

PROCESS.

Prepare the sample in the usual manner, being sure to use a 100-mesh sieve. Sample and weigh out very carefully on the ore scales, one gramme, if the ore be at all rich (say above 20 per cent), or five grammes if it be poor in copper (below 20 per cent).

Brush into a casserole, *i.e.*, a porcelain evaporating dish with a handle (fig. 91), and cover with a clock-glass (fig. 84), of slightly larger dimensions, and add 10 cubic centimetres of pure and concentrated nitric acid, by means of a 10 c.c. pipette (fig. 92).

Now place the casserole either on a sand-bath (a common tin plate holding some dry

---

* If accessible, consult a very interesting paper, entitled "Comparison of Various Methods of Copper Analysis," by Mr. W. E. C. Eustis of Boston, which was read at the August, 1882, meeting in Colorado of the American Institute of Mining Engineers, and is to be found among the published transactions of that society.

sand), or on a piece of wire gauze, supporting either on a ring-stand (fig. 97), and heat with a Bunsen burner (fig. 99), or alcohol lamp. Continue this heating some little time, then let cool. When cold, add, from another pipette, 5 c.c. of pure and concentrated sulphuric acid, and heat to boiling till no more *red* fumes are given off, but in their stead dense *white* vapors are delivered.

The red fumes are from the nitric acid, the excess of which we wish to get rid of, which is done by means of the sulphuric acid, and the white fumes show that the former acid is about gone. Let the casserole stand till cold.

Now add about 50 c.c. of distilled water, stir with a glass stirring rod, heat, and let stand till any undissolved matters have settled to the bottom of the casserole.

While this is doing, prepare for filtering, that is the separating of the dissolved copper (and other metals) from the undissolved silica, etc. Place in proper position a filter-stand (fig. 121), glass funnel (fig. 90), and

glass beaker (fig. 89). The filter-paper is
fitted by cutting a piece in a square, then
folding in half, diagonally, and then into quarters ; it will form a triangular figure, and if the
corners are cut off in a curved line, a circle
will be formed on spreading out. Upon opening the folded paper so that three thicknesses
come on one side and one on the other, a
filter is obtained, which is placed in the funnel
and wetted by means of the wash-bottle (fig.
83). This useful piece of apparatus is operated by simply blowing in at *a* ; a fine stream
of water at once issues from *b*, which can be
directed against any part of the funnel.

Filter the liquid in the casserole by holding
the glass-rod outside the lip of the vessel,
allowing the solution to run down the rod
into the funnel, till the latter is nearly full.
Repeat the operation until nearly all the solution has passed through the funnel, and only
the sediment, nearly dry, is left in the casserole.
Examine the residue, and if it is dark-colored,
it is best to repeat the treatment with acids.

Generally, however, once is enough. Finally wash the contents of the casserole into the funnel, which fill three or four times with water, which will be sufficient to wash out all the copper solution.

The residue on the filter-paper consists of silica and other substances insoluble in the acids used. It should contain no copper.

The filtrate, that is the filtered liquid, contains the copper as sulphate (with perhaps some nitrate), also it may be, iron, lead, etc., but these do no harm.

The next thing is to deposit the copper upon platinum. We may use a vessel entirely of platinum, or a copper dish lined with platinum, or a horseshoe shaped strip of platinum suspended in a glass beaker. In case the operator possesses the platinum or platinum-lined dish, clean it thoroughly by washing. If it is a new vessel, best rinse it first with some solution of caustic soda or potash to remove grease, then rub it gently with a little *fine* sand, thus giving the interior a surface favor-

able for deposition.   Be sure to wash off all
the soda or potash solution and sand, then
warm till it is perfectly dry.   When cool,
weigh carefully on the ore scales and note
the weight.   Pour the copper solution into
the platinum dish, using the glass-rod, which
rinse off with water into the dish; finally rinse
out the beaker with a little water into the
dish.

We now have a weighed dish containing
copper in solution from a known weight of
ore.   It remains to connect it with a battery,
which latter is now described.

FIG. 130.

Fig. 130 represents two cells of what is commonly known as the " Bunsen Carbon," which

form a battery powerful enough for our purpose. A quart size will be about right. It consists of a glass cell or jar to contain the dilute sulphuric acid, a cylinder of cast-zinc, of which the ends do not quite meet ; a porous earthenware cup, to hold nitric acid, and a rod of compressed carbon.

Prepare a mixture of strong sulphuric acid (ordinary commercial) and water in the proportion of one part of the former to ten of the latter, observing the precaution of pouring the acid into the water, never the reverse. Let the zincs stand in this acid solution for two or three minutes, then pour over them a little mercury, and rub with a piece of soft rag tied around a stick, till the entire surface, inner and outer, of the zincs, is coated with amalgam.

Put each cell properly together and fill the glass jars with the sulphuric acid mixture, just covering the tops of the zincs. Next, nearly fill the porous cells with concentrated (commercial) nitric acid. * See that the binding screws are filed bright, also the connecting wires, to make good contact. Arrange apparatus as follows :

The zinc of the first cell is to be united with the platinum dish by means of a coil of copper wire underneath the latter. A strip of platinum foil (cleaned with potash solution and sand) just dips into the solution of copper, and is connected to the carbon of the second cell by a copper wire. Another wire between the zinc of this latter cell and the carbon of the first, completes the circuit. Cover the dish with two pieces of window-glass, to prevent loss by spattering.

The copper at once begins to line the interior of the dish, and in from four to six hours the deposition will generally be complete. Time is often gained by starting the action at evening, and letting it run all night.

Prove the complete deposition of the cop-

---

* A bench of from two to six so-called "gravity" cells will do instead of the pair of Bunsen Carbons, and are more constant.

per by taking one drop of the solution and adding to it one drop of sulphuretted hydrogen water, mixing the two on a white surface (cover of a porcelain capsule). If no coloration ensues, the copper has all been thrown down ; if a black discoloration follows, then there is still copper in solution. In this latter case, continue the current till the test is negative.

In the former case, pour the contents of the dish into a clean beaker, rinse the dish, the under surfaces of the glass plates and the platinum strip, into the same beaker. On adding to the contents of this beaker an excess of aqua ammonia, no *blue* coloration should be seen.

Add a few drops of alcohol to the dish, rinse around and drain off. Set fire to the little r:maining in the dish, and when the latter is cool, weigh. The difference between this latter weight and the original weight of the dish is metallic copper. I give an example :

Weight of ore taken = 1 gramme.

GRAMMES.

Weight of platinum dish and copper = 56.408

"   "   "   "   empty = 55.659

"   "   copper from 1 gramme = 0.749

0.749 multiplied by 100, gives 74.9 per cent of metallic copper in the ore.

As before mentioned, instead of the platinum dish, which is quite expensive, a glass beaker can be used to contain the copper solution. A second platinum strip, on which is to be deposited the copper, must be used here. Dip this in the beaker and connect with a zinc element. The platinum strip in straight form connects with the other carbon as usual.

If the copper should form dark colored on the platinum, it is because the solution is too acid. *Nearly* neutralize with a little ammonia water, to counteract its bad effect. Too strong a current should also be avoided. I have given this process well in detail, but it will be found to be much easier learned than described. It is a very pretty and satisfactory method.

Volumetric analysis of copper ores.

If very many tests of copper ores, slags, mattes, etc., are to be made daily, it will be better to use the volumetric process, for while it is not quite so accurate as the battery process just described, it is much more rapid.

My thanks are hereby tendered to Mr. F. E. Fielding, of Virginia City, Nev., for many of the details embodied in the following description.

Preparation of standard solutions.

Solution of cyanide of potassium.—Dissolve 130.2 grammes of the pure salt in 1 litre (1,000 c.c.) of distilled water, and preserve from the light in a glass-stoppered bottle.

Solution of metallic copper. — Dissolve 1 gramme of pure copper in 25 c.c. nitric acid of 32° Beaumé (specific gravity 1.26, see page 211) in a covered beaker; then add 7 c.c. of dilute sulphuric acid. When dissolved and cool add caustic ammonia, or, better still, ammonium carbonate, in slight excess, or until the blue color is perceptible throughout. Di-

lute, when cool, with distilled water to exactly
1 litre.

Standardizing the cyanide solution.— Divide
the copper solution in two beakers, so that
exactly 500 c.c. shall be in each one, representing ½ gramme of pure copper.   Run in
the cyanide solution from a burette until a
faint violet color is produced.   Treat the second portion in the same manner, and take the
mean of the two results as the standard.
Multiply each result by two to bring it up to
the basis of 1 gramme of pure copper, representing 100 per cent.   The cyanide solution
should be re-standardized every day or so.

Preparation of the ore solution.

Weigh 2 grammes of the ore, or 5 grammes
if very poor in copper, place in a flask, beaker,
or casserole, moisten with about 7 c.c. sulphuric acid, then add 25 c.c. of nitric acid.
Digest at a gentle heat, or until the silicates
look white, showing that all the metals have
been extracted.   Boil until fumes of nitrous
acid are no longer evolved, cool, and add am-

monia or carbonate of ammonia in slight excess; allow the solution to become cool and the separated flakes of brown-red hydrated sesquioxide of iron (if present) to settle to the bottom of the vessel. If necessary, filter, and wash the filter paper thoroughly.

The analysis.

The standard solution of cyanide is now gradually added from the burette (constantly stirring the copper solution), until the blue color has entirely disappeared, and has been replaced by a faint violet tint, corresponding as nearly as possible to the shade produced in titrating the solution of pure copper in standardizing the cyanide solution. The number of c.c. used to produce this is now read off, and from it the percentage of copper in the ore is calculated. A single example will illustrate the simplicity of the calculations.

We will imagine it required 125 c.c. of the cyanide solution to neutralize the 1 gramme of pure copper representing 100 per cent., while it took but 75 c.c. for, say, 2 grammes of ore.

125 × 2=250 ; 75÷250=0.30 ; 0.30 × 100=30, or 30 per cent.

Hence the rule : Multiply the number of c.c. of cyanide solution used for the pure copper test by the number of grammes of ore taken. Divide the product *into* the number of c.c. employed on the ore test, and the quotient multiplied by 100 will give the percentage of copper in the ore.

When a sulphuretted ore is to be operated upon, it will, in a majority of cases, be completely oxidized by a mixture of sulphuric and nitric acids, but should any globules of sulphur remain, they may be taken out after the dilution of the solution, ignited, and the residue dissolved by nitric acid, and added to the original solution. (Or such ores may previously be roasted in the manner described in the section on gold and silver ores.) Some ores are best attacked by aqua regia (1 part nitric acid to three parts hydrochloric acid).

Owing to the influence of varying quantities of ammonia and of ammonium salts upon the decolorization of copper solutions by potassium cyanide, it is necessary that both the test solution originally prepared and the various copper solutions subsequently titrated should contain, as nearly as possible, equal amounts of ammonia.

Interferences, and their removal.

Zinc, nickel, and cobalt.—These, if present, render the analyses unreliable, so that, in such cases, the copper must be removed by precipitation from its solution. This may be done by placing a piece of iron or zinc in the solution — care being taken that nitric acid is not present. The precipitated metallic copper thus obtained is, after removal from the remaining solution of interfering metals, dissolved in the usual manner.* The copper may also be precipitated as sulphide by means of sulphuretted hydrogen (see page 165) in

---

* Instead of re-dissolving, it can be dried and weighed as metallic copper, giving approximately the percentage of this metal in the ore.

acid solution, or by solution of sodium thio-sulphate (hyposulphite of soda), and the sul-phide re-dissolved and estimated as described.

Manganese is not often found in copper ores in sufficient quantity to interfere. If present, it may be removed by adding to the ammoniacal solution sodium carbonate, and a few c.c. of bromine water, and boiling. When cool add the standard solution in the usual manner.

Arsenic does not interfere, excepting in the presence of iron, when it forms an arseniate soluble in ammonia, and gives rise to a brown-ish color in the liquid. It may be removed by adding magnesium sulphate (epsom salt) in excess. As soon as a precipitate is no longer formed, and the solution has recovered its characteristic blue color, run in the stand-ard solution.

Silver.—If this be present to any great ex-tent, it may be removed by adding a few drops of hydrochloric acid to the solution, and filtering *before* the addition of ammonia. (Or the hydrochloric acid may be added at same time as the nitric and sulphuric acids.)

Iron.—This, in the state of ferric hydrate, does not interfere chemically, but obscures changes of color by its being disseminated throughout the solution ; hence, it must be allowed to settle after each addition of the standard solution. It may be kept in solution by means of tartaric or citric acid.

If it is allowed to be thrown down by the ammonia, whether to be removed by filtration or not, it should be borne in mind that in the case of ores very rich in iron, the iron precipitate will retain copper, which cannot be dissolved out of it by either boiling water or ammoniacal water. Up to about five per cent of iron, no special pains need be taken, but above that amount, the copper ought to be first removed from the solution by sulphuretted hydrogen, then dissolved as usual.

Lead, bismuth, antimony, and magnesia do not interfere.

Lime in large quantity tends to confuse the

results ; it may be removed by addition of oxalate of ammonia.

The preceding statements regarding interferences are well borne out by a series of experiments carried on by Messrs. Torrey and Eaton, New York, and described in the *Engineering and Mining Journal.* New York, May 9, June 9 and 27, 1885. The results, in brief, were as follows :

Zinc : 3 per cent caused an (apparent) increase of $\frac{1}{2}$ per cent copper.

Arsenic : From 5 to 15 per cent did no harm.

Silver : 25 per cent caused an error of only $\frac{1}{10}$ per cent copper.

Iron : 30 per cent caused a loss of 3.71 per cent copper.

Lead : From 5 to 40 per cent had no injurious effect.

Bismuth : 20 per cent allowed the same approximation as the silver.

Consult the above article, also "On the Volumetric Determination of Copper by Means of Potassic Cyanide," by J. J. and C.

Beringer, *Chemical News*, Vol. XLVIII, No. 1,241, Sept. 7, 1883, pp. 111–113, and Sutton and Hart on volumetric analysis.

IV. AMALGAMATION ASSAY OR LABORATORY MILL RUN.

By M. G. Nixon, M.E.

The wet copper assay bears somewhat the relation to the fire copper assay that the fire gold assay does to the amalgamation gold assay.

In a certain sense, no one cares to know the ultimate amount of metal that an ore contains. What is desired in practice, is the *yield* under the most skilful treatment, and this information is approximately obtained by fire for copper, and the amalgamation process for gold.

There are those so practised in "panning," that from a "panful" of "pulp" they can very closely guess the yield by the number of "colors" and their size. Of course this method is not very popular, nor can it ever be,

Something more a matter of weighing, and less a matter of judgment and practice is required.

The amalgamation assay in its simplest form consists in "panning" a weighed amount of "pulp" with few or many drops of mercury, accordingly as the ore is poor or rich. The tailings are washed out as clean as may be, the pan is then placed over a fire to dry and then what remains of dirt and dust is blown out with the breath; the pan is again placed over the fire and the mercury volatilized, leaving the gold ("retort") ready for weighing. This process is quite largely followed by prospectors in some of our free-gold districts.

An improvement on the method just described consists in grinding the pulverized ore in a large iron mortar with which water and mercury are introduced, with the pestle. When the grinding is complete, the whole is washed into a pan to be collected and finished as before.

These methods are not recommended, but

may be resorted to when other apparatus can not be obtained.

The third method consists in grinding say ten or twenty pounds of ore in a laboratory "arrastre" by hand two hours or more, or, where possible, by power half as long. It is well to pass the ore through a 40-mesh sieve before placing it in the "arrastre." From two to four ounces of mercury are then squeezed through a piece of chamois skin, or blown through a tube the end of which is drawn out so as to make a pin-hole exit. Having put the pulp and mercury into the "arrastre" mortar, a piece of potassium cyanide as large as the end of one's little finger is dropped in, the grinder adjusted, enough water added to cover the ore, and the grinding performed. After it is finished, the grinder is first washed off into a collecting pan, then the mortar with its contents is treated in the same manner. The best way to collect the amalgam is to hold the pan under a running stream or water faucet, and very gently to stir it with the hand. The amalgam is then placed in chamois skin and squeezed so as to get rid of as much mercury as possible.

The residue is next placed in a small iron retort, and what remains of the mercury is driven off by heat gradually increased. Of course, for reasons of economy, it is well to condense the mercury; it may then be sold to mills or others, but neither that portion condensed nor that squeezed through the chamois should be used over again, since it is almost impossible to get rid of the last traces of gold. The "retort" is then to be scorified, cupelled, inquarted, etc., etc.*

The writer has saved 93 per cent of the fire

---

*An amalgam obtained as a result from either the preceding method, from panning with mercury, or from any other process, can be treated in the following manner, provided it is not in too great quantity.

Into a new scorifier (say 2¾ inches in diameter) introduce the amalgam after it has been separated from the free mercury by squeezing in a piece of chamois skin. On top of the scorifier place another of same size, *inverted*, having first bored through it a hole about ⅛ inch in diameter. By rubbing down a little the tops of the scorifiers, and painting their edges with a thick wash assay on the same ore that a mill at the time under his superintendence returned 89 per cent.

V. PAN TEST FOR GOLD.

("PANNING.")

The estimation of gold in ores in which the metal is in the *free state* is unreliable by

of ground chalk and water, all danger of loss of amalgam by its spirting through at the sides, is avoided.

Heat in muffle or furnace till the mercury has been driven off in vapor through the fine opening above, take out, let cool and remove the upper scorifier.

Now put the chamois skin on top of the residue in the scorifier and burn to ashes in muffle or furnace, remove a second time and cool.

Finally mix the residue and ashes with granulated lead and scorify. Re-scorify with more lead if the resulting button is brittle. Cupel in the usual manner and treat the bead obtained as gold bullion.

If it is not thought worth while to save the mercury, the fluid amalgam can be treated directly as described without first squeezing through a chamois skin, in which case the accompanying step of burning the latter, etc., is dispensed with. Heat very gradually. Even with this apparatus the mercury can be saved by attaching to the upper scorifier a small iron tube which bends over and dips into water.

The advantages of the above method are that it is simple, easy to operate, and that all the work (up to the cupellation) is done in one vessel, and so any liability to loss of gold in transferring from a retort, etc., is done away with. Furthermore, working with small quantities of amalgam, in even the smallest retort obtainable, is unsatisfactory.—W. L. B.

either the crucible or scorification process, owing to the impossibility of securing an average sample.

The ore, for supposition, may be of such value that even when put through a 100-mesh sieve one flake that would go through such a mesh could represent the amount of gold in two assay tons.

If then of two assay tons of ore of the above character, one is taken, it must run either nothing or double the true value of the ore.

Again, on low grade ores and with the charge most convenient to employ, the result or weighable button is so small that its estimation is liable to error.

Many ores containing small quantities of gold are frequently profitable to work, as in the case of placers and of large quartz ledges where the rock is soft and gold free. In such cases the assay report based upon the *small quantity* of ore used in the scorification, or even in the crucible assay, is unsatisfactory for this and the previous reasons.

Here, then, we resort to the pan test, for by it we can treat large amounts of ore, and the greater the quantity operated upon the more reliable the result.

The pan test is a process of concentration (doing on a small scale that which concentrators effect on the large), the product being either gold particles, or gold sulphurets, iron, sand, etc., depending on how far the process is carried.

The pan itself is a Russia sheet iron vessel of a shallow truncated conical shape (diameter about 16½ inches). That form sold by mining outfit establishments has been found most useful in practical operations. A round shallow wooden dish with its bottom sloping to a point, and technically known as a " Batea," is a useful modification (fig. 116, page 138; size of batea: diameter, 17 inches; depth, 1⅞ inches; thickness, ⅝ inch; angle of sides, 12°; material, Honduras mahogany). Each person will exercise his own choice after learning the operation. After the requisite skill has been acquired, a pan can be extemporized from almost any kind of dish, or a section of bullock's horn or an iron spoon may serve as substitutes.

The requisite amount of ore from 100 to 500 assay tons (5 to 25 pounds, or in French weights 3 to 15 kilos), depending upon its richness, is sampled, crushed and pulverized as directed in the chapter on gold and silver ore. The pulverization, however, need be carried no finer than to cause the ore to pass through a 40, 50 or 60 mesh sieve (the latter preferred). Weigh now the ore and put in the pan, which latter must be free from grease. Moisten and let it stand for a few moments in order that particles may not float off when the pan is put in water.

When wet, the whole pan and ore is gently sunk below the surface of a tank of water (a common wash tub will do nicely in the laboratory). A peculiar oscillatory motion or side vibration is commenced, though not enough to throw any particles of ore over the edges of the pan. The object of this is to settle the heavier particles (the free gold, heavy minerals, black sand, etc.), and have nothing on the surface but rock or quartz ; a little experience will teach the point. Then slightly incline the pan, and so wash it around as to carry the surface rock over the edge ; only a little at a time, however.

Level the pan and resettle as at first ; again incline and wash more over the edge. Keep up this operation, gradually getting more and more rock over the edge, and becoming more careful and washing more delicately as the process continues.

Toward the end of the operation, that is, when the rock is nearly gone, be careful to keep the ore under the surface of the water, as the gold might otherwise become dry and float off. Also make no sudden or unusual lurch, or the whole result may go off the pan. The above manipulation is far more difficult to describe than to perform after having once been acquired. Dry the residue.

If gold alone is obtained, that is, gold (or gold and silver) free from sulphurets, etc., it must be treated as an alloy, weighed, parted and weighed again, or cupelled with lead, weighed, parted and weighed; in both cases giving gold and silver.

If the panning is not carried to such a point as to get rid of all the rock, the concentration is all scorified with test lead (or run down in a crucible), cupelled, parted and weighed. In the case of an ore supposed to carry auriferous sulphurets it should be panned so far as can safely be done without losing metalliferous particles and the concentration treated as above described.

If the ore is quite poor, or a large quantity is desired to be worked, the panning can be carried on roughly and the successive concentrations finally panned together.

The results are based upon the amount of ore taken in the pan. If much of this work was to be done, a set of weights from 500 A. T. down (approximately accurate) would be very convenient and save calculation. The result would be as many times the number of ounces contained in the ore as the quantity of ore was more than one assay ton.

For example, the ore was supposed to be very poor and therefore:

> 500 A. T. were taken.
>
> Bead weighed 50 mgrms.
>
> ∴ 500 A. T. : 1 A. T. :: 50 mgrms. : $\frac{1}{10}$ mgrm.,

or the ore ran $\frac{1}{10}$ oz. Troy per ton.

If 100 A. T. had been taken and the same weight bead obtained, we would have:

> 100 A. T. : 1 A. T. :: 50 mgrms. : $\frac{1}{2}$ mgrm.,

or the ore would run $\frac{1}{2}$ oz. Troy per ton.

As an example of the calculation required *without* the large assay ton weights, I give the following:

Weight of panful of ore, $2\frac{1}{4}$ kilogrammes = 2,250,000 milligrammes.

Weight of bead obtained, gold 20 mgrs., silver 50 mgrs.,

$$\text{then } \frac{2,250,000 \text{ mgrms.}}{20 \text{ mgrms.}} :: \frac{29166}{x} \times = \tfrac{25}{100} \text{ oz.,}$$

$$\text{and } \frac{2,250,000 \text{ mgrms.}}{50 \text{ mgrms.}} :: \frac{29166}{x} \times = \tfrac{64}{100} \text{ oz.}$$

The free gold can be separated from the

sulphurets (if it be desired to determine how much of the gold is "free" and how much in the "sulphurets") by washing in an amalgamated pan. Such a vessel may be simply made by bending a piece of thin silver-plated copper (about 6 inches by 12 inches) so as to form curved edges on three sides, the silvered sides in. The side not turned up is one of the narrow ends. A little mercury (free from gold and silver) will quickly amalgamate the interior, and if the ore is washed carefully over this, most of the free gold will become amalgamated and stick to the pan. A piece of chamois skin made into a rubber will push the gold, which can be seen as little specks of amalgam, to the open edge of the pan and into a crucible. The mercury can be driven from the gold by heat.

No investigation has been made to determine if any silver is carried by the mercury to the assay from the pan, but if such be the fact, the result is still accurate for gold. If carefully performed the results ought to be above the yield from a stamp-mill with amalgamated plates.

A more common test than with the above silver-plated amalgamated copper pan, is, after having panned down, to drop a few globules of clean "quicksilver" (*i.e.*, mercury) into the pan and a little cyanide of potassium (to keep the mercury clean). Work up with a spatula till the mercury has taken up the free gold, then collect, and run off the mercury. Clean it and dissolve in nitric acid (for the gold only) or drive off the mercury in the muffle, weigh the residue of gold and silver, part and weigh gold.

The residue in the pan should then be assayed and the gold and silver (actual weight) determined. Suppose

Original weight of ore............$2\frac{1}{4}$ kilos.

Gold and silver after retorting......35 mgrms.

Gold after parting................15   "

Hence silver ....20 mgrms.

Gold in sulphurets................50 mgrms.

Silver   " 90   "

Then we have :

Free gold $\frac{19}{100}$ oz. per ton of original ore.

Silver in free gold $\frac{25}{100}$ oz. per ton of original ore.

Gold in sulphurets $\frac{64}{100}$ oz. per ton of original ore.

Silver in sulphurets $1\frac{16}{100}$ oz. per ton of original ore.

Total gold $\frac{83}{100}$ oz. per ton of original ore.

Total silver $1\frac{41}{100}$ oz. per ton of original ore.

There is a certain loss in panning, hence the results are not analytically accurate, but are close indications of the practical result of the working of gold ores in a mill with copper plates.*

VI. CHLORINATION ASSAY OF GOLD ORES.

If gold exists free in the gangue, that is, not combined with sulphur, arsenic or tellurium, it can be chlorinated directly without roasting.

*For the information comprised in the above article I am largely indebted to Mr. S. A. Reed of Irwin, Col., and Mr. Ray G. Coates of Chicago,

But sulphurets, arseniurets or tellurides must be first roasted and thoroughly at that.

The chlorination can be done in the laboratory on either a large or moderately large scale. For the former, operating say on 20 pounds ($\frac{1}{100}$ of a ton), consult the section in Kustel, entitled "Extraction of Gold from Sulphurets, Arseniurets or Quartz, by Chlorination," pp. 136–139.

For the latter grind up 5 to 8 ounces (or 5 to 10 A. T.), and, if necessary, roast in the usual manner. Use a frying pan for this purpose, and see that the sulphur is entirely driven out so that no smell (as of a burning match) is perceptible at the finish. Cool, grind in an iron mortar, and re-roast at a red heat.

When cold, reserve 1 A. T. of the ore for regular assay; the remainder is to be chlorinated in the apparatus herewith described. It consists of a flask, provided with a funnel tube for acid supply and delivery tube for the chlorine gas generated. The latter tube dips into a wash bottle containing water to wash the gas. From the latter the gas passes up into a separatory funnel containing the ore. The exit tube from the funnel may pass into a flue or the open air, or into a cylinder holding shavings moistened with alcohol.

Place in the flask a mixture of 3 parts of black oxide of manganese, 4 parts of common salt, and $4\frac{1}{2}$ parts of water, all well mixed.

Place the ore, which has been dampened with water, in the separatory funnel, having put in at the bottom a very little cotton to prevent the fine ore from stopping the passage of the gas.

Having now made all ready, pour down through the funnel tube $7\frac{1}{2}$ parts of sulphuric acid at intervals. After a time the flask is to be gently heated, that all the chlorine may be driven off.

Run the operation for about two hours, then disconnect the flask and let the funnel stand over night. Finally take out the upper cork and wash out the chloride of gold with distilled water.

To the solution in a beaker or tumbler add a few drops of hydrochloric (muriatic) acid, then some solution in water of sulphate of iron (green vitriol or copperas), stir with a glass rod, warm and let stand undisturbed until all the gold has been thrown down to the bottom and the liquid above is perfectly clear. Half of this liquid is drawn off with a syphon, the remainder containing the gold is filtered as usual, washing with warm water. Dry the filter, burn, scorify ashes and cupel, or cupel directly with sheet lead, weigh, etc. Compare result with unchlorinated sample. Consult Kustel as above, and Aaron's "Leaching Gold and Silver Ores," p. 90, on the "Working Test."

VII. CHLORINATION TEST FOR SILVER.

In smelting-works it is often necessary to test ores that have been subjected to chloridizing roasting, to ascertain the amounts of chloride of silver contained in them. Two assays are made of each ore.

Several pounds of the ore are taken from various portions of the entire lot, well mixed and sifted. From this, weigh out two charges of $\frac{1}{10}$ A. T. Scorify and cupel one charge in the usual manner.

The other charge is brushed into a filter paper held in a glass funnel, and over it pour a *warm* solution of hyposulphite of soda (six or eight ounces in a quart of water), which rapidly dissolves the chloride of silver from the ore. Continue this treatment until a small portion of the filtered liquid contained in a test-tube, darkens but slightly and does not lose its transparency upon the addition of a few drops of a solution in water of sulphide of sodium.

Wash the mass in the filter with warm water, remove filter and all, dry and burn in scorifier in muffle, at a low heat, mix ashes with lead, scorify and cupel as usual.

(The hyposulphite solution dissolves out sulphate of silver as well as the chloride. If, as is sometimes the case, it is desired to know the amount of sulphate present, leach a third charge with warm water, which will take out the sulphate, but will not touch the chloride or any unacted-upon ore. Scorify and cupel the residue as directed.)

The difference between the two cupellations shows the amount of silver which has been changed into the state of a chloride. Thus:

1st charge ran 180 oz. per ton.
2d charge ran 10 oz. per ton.
Hence, 180—10=170 oz. of chloridized pulp.
To obtain percentage:
180 : 170 :: 100 : x = 94.4 per cent.

"If there should be gold in the ore, this must be subtracted from both assays, because, although the amount of gold would be equal, the chlorination result, as it should be, must come out higher after the gold is subtracted." (Kustel.)

VIII.—THE ASSAY OF GOLD AND SILVER BULLION.

The larger portion of this article is taken from the Fourth Annual Report of the State

Mineralogist of California for the year ending May 15, 1884, by the kind permission of Mr. Henry G. Hanks, State Mineralogist. His very able and complete paper could hardly be improved, hence the extracts are given almost *verbatim*, as many will not succeed in obtaining copies of the report, the edition of which is already exhausted.

The remainder of the description is made up partially from information given me personally by Mr. F. E. Fielding, assayer of the Consolidated Virginia and Consolidated California Mines, of the Comstock Lode, Virginia City, Nevada, and partially from other sources, together with a little as the results of personal experience.

Bullion, or the precious metals in bars more or less impure, is of very varying composition, running down from that which is mostly gold, through mixtures where the gold and silver are more equally divided, and where the silver predominates (all containing various impurities or base metals, as lead, copper, etc.), to base bullion, or that in which the lead is in excess of all the others. The latter kind is treated of in another portion of the appendix. Here I consider only those bullions in which the gold and silver form the greater amount.

Absolutely accurate assays of gold and silver bullion require care, skill, and first-class apparatus. The skill may soon be acquired by practice, but the apparatus must not only be of the very best quality, but must be kept in the most perfect state of adjustment. It is not enough to purchase chemicals which are marked "pure," or a balance supposed to be accurate. The chemicals must be tested, and the accuracy and adjustment of the balance and weights verified, before correct results can be certain.

The process of assaying gold and silver bullion is divided into several operations, as follows : Melting and refining the crude bullion, and casting the bar, cutting the assay chips, or otherwise preparing the assay samples, the preliminary assay, the assay proper, calculating the results, weighing the bar, and stamping the fineness and value upon it.

Melting, refining, and casting the crude bullion.— For melting, a wind furnace is best, but a good coal stove, such as used in offices, will answer the purpose if the amount operated upon is small.

The wind furnace is a square box of fire-bricks, built in the form of a cube of three-foot face, with an opening in the centre of the upper face. The fire-box is about a foot square, and fourteen inches deep, provided with an ash pit, movable grate, bars, and sliding cast-iron cover. The flue should be a horizontal opening, about three by six inches, near the top of the fire-box, and connected with a chimney at least thirty feet high, to insure a good draft. The furnace can be built by any bricklayer of ordinary skill and judgment. No mortar should be used in laying the fire-brick, but good clay, mixed with a portion of coarse sand, substituted.

The bullion is generally melted in a black

lead crucible. Before such a crucible can be safely used, it must be annealed. Were this neglected, and it should be placed in the fire without this precaution, it would soon fly to pieces. This is caused by the water it contains being converted into steam; and the structure of the material being such that the steam cannot make its escape, destruction of the crucible follows. It is best to commence annealing the crucible some time before it is wanted. It should be set near the hot furnace for several days, and turned occasionally. When the fire is nearly spent, it may be placed, rim downward, upon the hot sand, generally placed on top of the furnace. A day or two of such treatment will make it safe to hold it over the open furnace by the aid of the crucible tongs or poker. After it has been frequently turned, and is hotter than boiling water, it is safe to place it, rim downward, upon the burning coals. After the rim is red hot, all danger is passed, and it may be

turned, and placed in position for the recep-
tion of the gold.

If the fuel is charcoal, it will be best not to
use small pieces, or, at least, not coal dust.
Pieces the size of an egg, or larger, will make
the best fire.   When the crucible becomes
red hot, a long piece of quarter-inch gas pipe
is used to blow out any dust or ashes that
may have fallen into it.   A cover is then
placed on the crucible, and lumps of coal built
up around it with a long pair of cupel tongs.
When the crucible has attained a full, red
heat, one or two spoonfuls of borax, wrapped
in paper, are placed in it, using the cupel
tongs.   When the borax has melted, a small
quantity of the bullion, also wrapped in paper,
is placed in the crucible in the same manner.
Several portions may be thus added, according to the size of the crucible.   A fresh supply of charcoal must be built up around the
crucible when required, the cover having been
previously replaced.   When the bullion has
melted down, more is added in the same man-

ner, until the crucible has received all that is to constitute the bar. In the meantime, the ingot mould, in which it is intended to cast the gold, must be made smooth and clean inside. This is best done by rubbing with sandpaper and oil, or with a dry piece of pumice stone. It is then wiped dry and clean with a rag, oiled slightly, and placed on the edge of the furnace in such a position that it may become quite hot; not so hot, however, as to approach redness, nor to cause the oil to burn.

When the bullion is in a fluid state in the crucible, the mould must be placed on a level surface, and oil poured into it. To make a clean bar, it will be found best to use considerable oil — sufficient to cover the bottom of the mould to the depth of at least one-fourth of an inch. The mould should be turned in such a manner as to allow the oil to flow to all parts of its interior, and then placed again level, and in the position it is to occupy while casting the bullion. If the latter is clean, and the quantity less than fifty ounces, it is best not to attempt to skim it. Two spoonfuls of nitrate of potash may be added, and one of carbonate of soda, and the whole allowed to melt and flow over the surface of the melted metal. When very hot, and the slag perfectly fluid, the crucible is lifted from the furnace, and with a bold and steady hand, its contents are poured into the mould, the crucible being held for a little time in an inverted position, to allow the last portion of metal to flow from it. The oil inflames, and remains burning on the slag, which flows. evenly on the surface of the bullion. If the mould is clean, and of the right temperature, and if sufficient oil is used, a clean bar will result. A little practice will enable the operator to hit the exact conditions. The oil used should be a cheap animal oil ; common whale oil answers every purpose ; lard oil is also well suited ; coal oil is too inflammable, as well as dangerous, and should never be used. When cold, the bar falls easily from the mould. A slight tap with a hammer separates the slag, and the bar may be cleaned with water and nitric acid, or, if necessary, with sand and a suitable brush. A good plan is to place the bar in the furnace until it becomes nearly red hot, and then to quench it suddenly in water. This will be unnecessary if proper precautions have been observed in preparing the mould.

When the bullion is very impure — which is the case when in the form of retorted amalgam which has not been properly cleaned — a different method of treatment should be adopted. A large-sized crucible will be required. Three or four times the amount of flux must be put in, with the addition of a spoonful of carbonate of potash. A skimmer must be prepared by forming the end of a large wire, about the size of a common lead pencil, into a spiral about an inch and a half in diameter, and bending it so that when the skimmer is let down vertically into the crucible the spiral will lie flat upon the surface of its contents. A bucket of water is set near the furnace, and when the slag has become fluid, and it is beyond question that the bullion has become perfectly melted, the skimmer is touched to the slag and gently moved from side to side; a portion of the slag adheres to the iron, the skimmer is removed and plunged into the water, and immediately replaced in the crucible; an additional portion attaches itself to the skimmer, which is again quenched in water. This is repeated until a large portion of the slag is removed, and a new charge of flux, consisting, this time, of borax and nitrate of potash, is allowed to fuse upon the surface of the bullion. The first flux is removed from the skimmer by a slight blow with a hammer, and the crucible is skimmed with it as before. This must be repeated until all iron and other impurities have been removed, and the surface of the molten metal appears, when exposed, clean and reflective as a mirror. It may then be poured into the mould, as described before. Care should be taken not to dip the wet skimmer beneath the surface of the bullion, or an explosion will take place.

In large meltings it is customary always to skim the bullion before pouring, and so far to remove the slag that any remaining portion may be left on the sides of the crucible, and the metal only allowed to flow into the mould. This requires some skill and considerable practice. As it is imperative that the bar should be homogeneous to insure a correct assay, it is usual to mix the melted metals thoroughly before pouring. This is done in the large way by stirring just before lifting from the furnace. It may be done with an iron rod, with a piece of black lead held with the tongs, or with a clay stirrer made specially for that purpose, in which case it will be necessary to allow it to remain in the crucible until it has acquired the temperature of the fused metal ; otherwise, a portion of the bullion may attach itself to the stirrer, and be removed with it. In small meltings it will be found sufficient to mix the bullion by giving the crucible a rotary motion while holding it with the tongs just previous to pouring. This must be done so quickly that the crucible has no time to cool. For very small fusions it is best to use a small Hessian crucible, and, when the bullion is melted with plenty of flux, to set it aside to cool, and then break the crucible, and separate the pieces of crucible and portions of slag by slight blows of a hammer on the edges of the button. It is very difficult to pour small quantities of gold without loss from portions remaining on the sides of the crucible.

Preparation of samples for assay.— When the bar is clean, a small portion must be taken from different parts for assay. This is done in several ways, very frequently by cutting pieces from opposite corners or edges with a cold-chisel or hollow punch, but this is extremely clumsy, and in every way inconvenient. If the bar is brittle, a much larger piece may break off with the chip than is required. If the proper-sized chip is cut off successfully, it is likely to fly away and be lost. A second way of sampling is to bore into the bar, top and bottom, with a small drill. This may be done in a lathe or by means of a ratchet drill. The bar should be placed in a clean, copper pan, so that no loss may occur ; the surface borings, resulting from the first revolutions of the drill, should be rejected. Those that follow, to the extent of a little more than one gramme, are to be placed in a suitable vessel, and carefully preserved for assay, each lot separate. Before cutting or boring the bar the number of the assay should be stamped upon it, and the same number placed with the clippings or borings. This number should represent the bar through every stage of the assay by which its value is ascertained. Some assayers stamp the initial of their name on the cut faces, so that no portion can be removed after it leaves their hands. A third manner of sampling is that by "granulation." While the bullion is still in the melted state in the crucible, but is already refined, it is well stirred, and two samples are scooped up with a small ladle, one from the bottom of the crucible, and the other from the top. Each ladleful is poured, slowly and carefully, and in a narrow stream, into a clean copper basin containing warm water, which is rotated quietly by means of a broom or paddle. Keep the resulting granulated metal from top and bottom of the crucible apart, drying each lot.

The selected pieces of the granulations, or the chips cut from the bars, are flattened on the anvil and passed through the rolls until thin enough to be readily cut by the snip-shears. (Fig. 102.) They, or the borings, if borings are taken, are now ready for weighing.

Weights and weighing.—For the bullion assay a special set of weights known as "gold weights" is used. As the basis of. the bullion assay is 1,000 parts, so the unit of the "gold weight" system is a 1,000 piece, from which the weights range down to a $\frac{1}{10000}$ piece. The actual weight of each piece in this system is one-half of its corresponding piece in the French or metric system; thus the 1,000 piece weighs actually 500 milligrammes, and so on. Gold and silver in bullion are always reported in thousandths; that is, in parts of one thousand, taken, as before stated, as the standard of the assay; hence the use of the weights described. By difference between the fineness and the 1,000, we learn the number of parts of the base metal contained in the bullion. Thus, if 1,000 parts of a bullion, after treatment, weighs 900 parts, it has a "fineness" of 900, or it is 900 "fine," and the base metal is 100. In weighing, always remember that the rider, or index needle, when marking tenths for gold, is to be multiplied by two, as each mark on the beam or index plate only represents $\frac{1}{10}$ in the metric system, while it should be $\frac{2}{10}$ for the "gold weight" system. Half-tenths are always reported in the tables for gold, but not in those for silver.

As shown from the preceding, the metric weights can be employed if the "gold weights" are not obtainable.

The method of weighing is conducted as follows: The assayer seats himself before the balance, having the clippings or borings in a convenient position inside the case. A 1,000 piece (or half a gramme weight) is placed in the right-hand pan of the balance, and portions of the clippings or borings in the other until nearly correct, but the bullion should be in excess. The largest piece is then removed by the aid of a pair of pincers, and a small corner cut off with the shears. This done once or twice will nearly balance the pans; but by touching the piece of bullion selected against a clean file, still more minute portions can be removed. By careful manipulation nearly the exact point will soon be obtained; but with the greatest care, if the balance is delicate, it will be found nearly impossible to adjust the weight so perfectly that the index needle will not point either one side or the other of the zero. In such a case, it will be necessary to make a memorandum of the error, and mark it with the number of the assay, and in weighing the cornet, to take the same reading of the index needle.

Preliminary Assay.—It has been found that silver cannot be dissolved out of an alloy of that metal with gold, unless the proportion of silver is at least two and one-half times that of the gold. If a larger proportion is present, the gold is left after the extraction of the silver in the form of a powder, and cannot be dried and weighed without danger of mechanical loss. If less, the gold protects the silver, and the action of the acid ceases, while some of the silver remains undissolved. An alloy of three parts of silver to one of gold was formerly taken, from which the terms quartation and inquartation come; but of late years the above proportions have been found to be best.

In order, then, to form such an alloy, the assayer should know the amount of silver in the bullion, that he may consider it in adding silver to a bullion mostly gold, or gold to a bullion nearly all silver. The following imaginary and reversed bullions will illustrate this point plainly:

BULLION NO. 1.

200 parts gold need 80 parts more to make 280 parts.
700   "   silver $= 2\frac{1}{2}$ times the 280 parts gold.
100   "   base metal.
_____
1,000   "

BULLION NO. 2.

700 parts gold
200   "   silver need 1,550 parts more to make 1,750 parts $=$
              $2\frac{1}{2}$ times the 700 parts gold.
100   "   base metal.
_____
1,000   "

The proportions of silver and gold in the bullion in question are ascertained in several ways. First, by merely looking at the bullion, having had previously such experience with similar alloys that the knowledge becomes almost instinctive. Secondly, by direct comparison with slips of known "fineness." Thirdly, by use of the touchstone and needles.

(For the methods of making and using the latter, see the article in Mr. Hanks' Report.) Fourthly, by a preliminary fire assay. This is by far the best way, and should always be done where great accuracy is required.

Take the 1,000 parts of the bullion already weighed out, wrap in a piece of pure sheet lead of about 2 grammes (30 grains or so) in weight, cupel and weigh the resulting bead. If it seems mostly silver, roll it out, boil in nitric acid of 32° Beaumé (1.26 sp. gr.), decant, wash, dry, ignite, and weigh the gold. If the bead appears very yellow, add three or four times as much pure silver, wrap up together in sheet lead, recupel, part as above, and weigh the gold. The latter may come down as a fine powder, and a slight loss occur in washing, but the results will be sufficiently accurate for the purpose.

From the data obtained above, it will be easy to make up the proper alloy for the regular assay and the proof centre, as shown by the following simple examples:

Example No. 1.

Suppose the 1,000 parts of bullion taken weigh, after cupellation, 520.5 parts, then 479.5 parts (1,000 — 520.5) are base metal.

The gold, after parting, weighs 15 parts.

These figures enable us to prepare the assay proper and the proof centre, as hereby shown:

Gold and silver bullion taken.......... 1,000.0 parts.
   "   "   "   after cupellation....... 520.5   "
          Leaving base metal........... 479.5   "
Gold and silver after cupellation....... 520.5 parts.
  " after parting................... 15.0   "
         Silver in the bullion .... 505.5   "
              505.5 × 2.5 = 202.2 parts.
Deduct gold already in the bullion..... 15.0   "
        Leaving gold to add........... 187.2   "

Hence:

Bullion		composed of		*Proof centre.*
take		505.5 parts silver	=	505.5 parts silver.
1,000.0 parts		479.5 " base	=	479.5 " copper.
		15.0 " gold		
187.2 "		gold to add	=	202.2 " gold.
1,187.2 "		total.		1,187.2 " total

Example No. 2.

Gold and silver bullion taken .... 1,000.0 parts.

" " " after cupellation....... 764.4 "

Leaving base metal............. 235.6 "

Gold and silver after cupellation....... 764.4 parts.

" after parting .... 602.1 "

Silver in the bullion .... 162.3 "

$$602.1 \times 2.5 = 1,505.2 \text{ parts.}$$

Deduct silver already in the bullion .... 162.3 "

Leaving silver to add .... 1,342.9 "

Hence:

Bullion	composed of		*Proof centre.*
take	602.1 parts gold	=	602.1 parts gold.
1,000.0 pts.	235.6 " base	=	235.6 " copper.
	162.3 " silver		
1,342.9 " silver to add		=	1,505.2 " silver.
2,342.9 " total.			2,342.9 " total.

The silver, gold, and copper added are, of course, pure.

The Assay Proper.— Two assays of a large bar are always made, the samples being taken from the top and bottom. A proof centre or "check" is always run through with each set of assays, and consists of pure gold, pure silver, and pure copper in such proportions as shall correspond as nearly as possible to those of the bullion under trial, as shown in the above examples.

The proof cupel is always placed in the centre of the muffle, those for the bullions at each side, the object being to correct the loss by volatilization.

The loss in weight of the proof centre bead is to be added to each weight of the bullion bead.

Make the weighings of the duplicate bullions, and the gold or silver that is to go with each, and also the three metals or so for the proof centre. Wrap each lot in a piece of pure sheet lead, and squeeze all into a bullet.

According to the difference in fineness of the bullions are the amounts of pure sheet lead varied, and the latter must contain no gold whatever. Its purity being established, it is easily prepared by rolling out to a uniform thickness (about $\frac{1}{64}$ inch or so), and about $1\frac{3}{4}$ inches wide. The weight of so many inches of the lead is determined, and the rest of the samples are cut from measurement of the one weighed. These pieces are known as "lead cornucopias," and should always be prepared by the assayer himself, and kept on hand in sufficient quantity.

If too much lead is used in the cupellation of the buttons, the loss in precious metals is increased by the greater length of time required for the cupellation. On the other hand, if there is a deficiency in lead, the beads are ill-shaped, and are liable to contain some of the base metals. If a large amount of copper is in the bullion, the lead must be increased. "Long experience has proved that silver opposes the oxidation of copper by its affinity, so that it is necessary to add a larger amount of lead in proportion to the quantity of silver present." (Mitchell.)

In cupelling bullion say from 980 to 1,000 fine, 30 grains (about 2 grammes) of lead are best used, and for very base bullion, and where the base is mostly copper, 100 to 130 grains (6.5 to 8.5 grammes, nearly) are generally used. Mark appropriately with ruddle three good cupels, place them in the muffle, the proof cupel between the other two, as directed before, and when hot, a piece of pure lead, weighing 3 grammes (about 45 grains), is placed in each. The leads will soon melt and begin to "drive"; that is, begin to be absorbed by the cupel; the assays are then to be added, using the cupel tongs. When perfectly melted, the cupels are drawn forward to that point in the muffle which experience has shown to the assayer that cupellation progresses most successfully. When the cupellation is finished, and the buttons have assumed a brilliant metallic lustre, they are removed, hammered slightly on their edges on a clean anvil, the last blow being given near one edge, to make that part thinner, in order to facilitate the rolling process which follows, and examined with a magnifying glass to see that all bone ash has been removed. Weigh and note results. The two bullion beads should weigh exactly alike; if this should not be the case, the heavier one must be examined carefully, to see if any particle of bone ash may have been overlooked. If this should fail, there is no recourse but to make another assay, which should agree with one of the first. Generally, if care is used, the first pair will agree. Then add to the weight of the assay beads proper the loss sustained by the bead of the proof centre. Next the beads should be annealed; which can be done in the muffle, if still hot, or upon charcoal, with the flame of a spirit lamp urged with a blowpipe. After cooling, they are passed through the rolls, being drawn into ribbons about $2\frac{1}{4}$ inches in length. The proper letter or number is stamped on the end of each slip, somewhat deeply, and all are then re-annealed. Each slip is then rolled up into a spiral form upon a glass rod or lead pencil, commencing at that end of the slip which is not stamped. A slight pinch, or reverse bend, after the rod is removed, will prevent their unrolling. The "cornets," so prepared, are then ready for treatment with acid, after which step the letters or numbers stamped upon them will be as distinctly seen as they were before.

Introduce each cornet into a separate parting flask or matrass, and add 1 fluid ounce (29½ c.c.) of pure nitric acid of 21° Beaumé (sp. gr. 1.16), place on the sand bath which acts as cover of the furnace, or on a small sand bath supported on the ring of a retort stand over a spirit lamp or gas burner, and boil until no more red fumes are evolved (say ten minutes). Just at the point of coming to a boil add one or two "pepper carbons" (made by heating whole or unground pepper beans to carbonization on an iron shovel or pan). They prevent spirting or bumping.

A folded piece of paper, or a pair of wooden tongs, is used to lift the flasks, and the acid decanted carefully into some convenient

vessel kept to receive it, as the silver is valuable, and may be recovered when a sufficient
quantity has accumulated. The same quantity of 32° Beaumé acid (1.26 sp. gr.) is then
poured into each flask, and, being placed on
the sand bath, again boiled for ten minutes,
with the "pepper carbons" as usual. After
this, the acid is poured off, and each flask is
filled up with distilled water, gently rotated,
and the water decanted and thrown away.
Repeat the washing, and finally fill, for the
third time, the flasks with distilled water, this
time quite to the brim. Over the mouth of
each flask an annealing cup is placed, mouth
downward, like a cap, and the flask and cup
inverted together. By these means the cornet
is deposited gently, and without loss or injury
in the cup. The flask is then gently raised
until on a level with the edge of the cup, when
with a quick side motion the flask is removed,
the water from it, of course, running to waste.
The water in the cup is poured out carefully,
and the cup and its cornet are heated, at first

gently on the sand bath till all the moisture
has been driven off, then in the muffle to redness, making the third time of annealing.
Take out and let cool.   The gold has recovered its natural color, and is firm enough to
be handled with pincers.   It must next be
weighed accurately, using the "gold weights"
(or the gramme weights and multiplying by
two), and noting any memorandum regarding
the position of the index in weighing out the
bullion in the first operation.   The weight of
the cornet in parts of the 1,000 piece (or in
half milligrammes) will represent the fineness
of gold in the bar, expressed, as before, in
thousandths.

A small amount of silver will always remain
in the cornet, no matter how carefully the
manipulations may have been conducted.
This *surcharge*, so called, must be deducted
from the weight of the gold by subtracting
from it one-half, one, or two thousandths (or
"points") accordingly as the fineness ranges
from 500 to 900, or as experience indicates.

Weighing the Bar.—The next step is to ascertain the weight of the bar in troy ounces and decimals. This must be done with the greatest accuracy. A good bullion balance is much to be desired; but a bar can be weighed on a defective balance if it is sufficiently delicate to turn distinctly with the hundredth part of a troy ounce. This method of weighing is called counterpoising, and is conducted as follows:

The beam must first be brought to a level by putting sand, small shot, or other convenient weights into the lighter pan. When in perfect equilibrium, a small weight is placed in one of the pans to test the delicacy of the movement, and if satisfactory, the bar is laid in one pan, and the equilibrium restored by putting any convenient substance, as sand, into the other. The bar is then removed, and ounce weights put in its place, which will be the exact weight of the bar, all errors of the apparatus being corrected by counter-poising, which will be evident to the reader without further explanation. Of course the ounce weights must be proved by experiment to be correct among themselves.

It is sometimes impossible to obtain troy ounce weights, in which case avoirdupois may be used. The same rule as to accuracy applies equally to them. Each pound equals 14.5833 troy ounces. An excess of even pounds must be made with ounces and decimals, which can be prepared by any person of moderate mechanical skill. The value of an avoirdupois ounce is 0.911458 ounce troy, or one-sixteenth of a pound. To make the calculation, it is only necessary to multiply pounds by the former and ounces by the latter factor, and add the two together. The following table may be used to facilitate the calculation:

Avoirdupois.		Troy Ounces.	Avoirdupois.		Troy Ounces.
1 ounce	=	0.911458	13 ounces	=	11.848958
2 ounces	=	1.822916	14 "	=	12.760416
3 "	=	2.734374	15 "	=	13.671874
4 "	=	3.645833	1 pound	=	14.583333
5 "	=	4.557291	2 pounds	=	29.166666
6 "	=	5.468749	3 "	=	43.749999
7 "	=	6.380208	4 "	=	58.333333
8 "	=	7.291666	5 "	=	72.916666
9 "	=	8.203124	6 "	=	87.499999
10 "	=	9.114583	7 "	=	102.083333
11 "	=	10.026041	8 "	=	116.666666
12 "	=	10.937499	9 "	=	131.249999

Suppose the bar to weigh twelve pounds and nine ounces; set the figures down thus:

```
  10.  pounds.
   2.  pounds.
  .9   ounces.
  ____
  12.9
```

Look for 10 pounds in the table, which will be the same as 1 pound with the decimal point moved one place to the right, 145.833; opposite 2 pounds will be found 29.166; 9 ounces will be found to be 8.203, which are to be added as follows:

10.	pounds................	145.833
2.	pounds................	29.166
.9	ounces	8.203

12.9	weight of the bar.	183.202 troy ounces.

When decimals of an ounce are calculated, the values may be taken from the first column of the table. Suppose the decimal to be .7, or $\frac{7}{10}$, move the decimal point in the seventh line one place to the left, and the result will will be .6380208, which is to be added to the sum of pounds and ounces.

The above method of weighing is sometimes convenient in isolated mining localities, where no accurate bullion balance or large sets of Troy weights can be obtained.

A table having been given to calculate Troy ounces from avoirdupois pounds, the following table has been prepared to reverse the operation, and it will in many cases be found convenient:

TABLE FOR CHANGING TROY OUNCES TO POUNDS AND
DECIMALS AVOIRDUPOIS.

Troy Ounces.	Pounds Avoirdupois.	Troy Ounces.	Pounds Avoirdupois.
1	.06857	6	.41142
2	.13714	7	.47999
3	.20571	8	.54856
4	.27428	9	.61713
5	.34285		

Gold is always estimated in troy ounces and decimals. A convenient set of weights may be constructed as follows :

Ounces.	Decimals.	Ounces.	Decimals.
500	0.500	10	0.010
300	0.300	10	0.010
200	0.200	5	0.005
100	0.106	2	0.002
50	0.050	2	0.002
20	0.020	1	0.001

Estimation of the Value of the Bar.— Suppose the total fineness (silver and gold) to be 900, and the fineness of gold as found by assay to be 100 ; by subtracting the latter from the former the fineness will be found to be 800. Now, as one ounce of pure gold is worth $20.6718, one one-thousandth will be worth $0.0206718; therefore, an ounce of alloy, containing 100 parts of pure gold, would be worth $0.02060718×100, or $2.06718. The last three decimals may be disregarded, unless the bar is very large.

The value of the silver is obtained in the same way. An ounce of pure silver is worth $1.2929, and one one-thousandth equals $0.0012929. This multiplied by the fineness of silver as found, would give the value of the silver in each ounce of the bar.

Multiplication may be avoided, and the calculations facilitated, by the employment of the following table :

TABLE FOR DETERMINING THE VALUE OF GOLD AND
SILVER BULLION.

Fineness.	Gold.	Fineness.	Silver.
.000½....	.010335917312	.000½ ...	.000646464646
.001	.020671834625	.001	.001292929292
.002	.041343669250	.002	.002585858584
.003	.062015503875	.003	.003878787876
.004	.082687338500	.004	.005171717168
.005	.103359173125	.005	.006464646460
.006	.124031007750	.006	.007757575752
.007	.144702842375	.007	.009050505044
.008	.165374677000	.008	.010343434336
.009	.186046511625	.009	.011636363628

The manner of using this table is the same
as a similar one described :

GOLD.

100 same as 001, decimal two places right $= \$2.06718$
$=$ value of gold per ounce.

SILVER.

800 same as 008, decimal two places right $= \$1.03434$
$=$ value of silver per ounce.

Value of gold per ounce...................$\$2.06718$
    " silver " 1.03434

Total value per ounce...................$\$3.10152$

These results, multiplied by the number of ounces and decimals of an ounce the bar weighs, would be its value in dollars and cents. Suppose the bar weighed 1,540.6 ounces, then —

$$\$2.06718 \times 1,540.6 = \quad \$3,184.69$$
$$1.03434 \times 1,540.6 = \quad 1,593.50$$

Total value of the bar = $\$4,778.19$

Stamping the Bar.—The assays being completed, the bar weighed, the calculations made, and values ascertained, there remains only to stamp the bar with the proper steel dies, giving the following data, which must be impressed in the bar before it can be sold : Number of the bar (which is the number of the assay also); name of assayer; the total weight of the bar, given in Troy ounces and decimals; fineness of gold; fineness of silver; total value of the bar in dollars and cents ; date.

IX. THE ASSAY OF BASE BULLION.

The uncertainty in the assay of base bullion lies, not in the determination of the amounts of gold and silver present, but in the difficulty of obtaining an average sample.

This question has given rise to an amicable discussion in the columns of the *Engineering and Mining Journal*, between various parties interested (issues of May 20, June 3, July 1, and September 9, 1882), eliciting some valuable information, which I purpose to reproduce herewith.

A base bullion may contain lead, silver, gold, copper, arsenic, antimony, and perhaps other metals, and sulphur. When this is melted and cooled, it tends to form alloys of varying degrees of fusibility, which with the dross or scum (a mixture of oxides, sulphides, etc.) make a pig or bar, from which it is not an easy matter to select a fair sample for assay.

In many smelting establishments the surface of the melted bullion is skimmed, and the clear lead ladled into the mould, till the latter is filled to within an inch of the top, and when it has solidified, the mould is filled

completely. There results then a nice-look-
ing bar, composed of good lead above and
below, with much dross in the centre. This
would not matter so much if an equal portion
of the dross could be gotten at for assay; but
there's the rub. The ordinary way of chip-
ping the top and bottom of the bar does no
good, since it seldom cuts deeper than $\frac{1}{8}$ inch
below the surface. Even a punch cutting a
chip $\frac{3}{4}$ inch deep does not solve the problem,
for it will not reach the dross when cast in the
middle of the bar.

Mr. L. S. Austin, in the issue of September 9 of the journal quoted, suggests a
method which seems to meet the requirement.
"It consists in the use of the punch which I
have already described [June 3d issue], and
which takes a chip of about $\frac{1}{8}$ inch in diameter and uniform in thickness. It is driven
clear in to one-half the depth of the bar by
the use of a sledge. The bar being, say, four
inches in depth, a chip a little over two inches
long is then taken both from top and bottom

of the bar. The chip is then slipped into a hole bored two inches deep into a block, and the projecting lower end trimmed off with shears to the exact length of two inches. Each chip represents, consequently, one-half the bar, its companion representing the other half; moreover, each chip is of the same weight. Thus each bar is represented according to its relative weight and to its entire depth."

Having obtained these chips, they are next melted, and poured into a small mould. Take this sample bar, cut slices across, each slice being a section of the bar. Cut from these slices $\frac{1}{2}$ A. T. for assay. By running five of these $\frac{1}{2}$ A. T. assays, and uniting the silver beads obtained, for parting, the gold present can be accurately determined.

Cupel the samples, "feathering" the cupels. Brittle or hard bullion can be scorified first, if necessary.

Consult the numbers of the journal referred to.

X. QUALITATIVE TESTS.

I have thought it a good plan to give a few simple wet tests for some of the metals, and acids united with them, as found in ores.

Ordinarily these tests work better on the powdered ore, though sometimes, as will be mentioned, the original rock can be directly treated.

Carbonates.—Place a drop of any strong acid upon the suspected rock ; if effervescence (or boiling up) ensues, unaccompanied by any odor, it contains carbonates. This test does not always show well with small quantities of carbonates ; try then some of the powdered ore with acid in a test-tube. To confirm the presence of carbonic acid, suspend in the test-tube a glass rod that has previously been dipped in lime-water ; the drop on the rod should become turbid or milky, owing to the formation of carbonate of lime.

Place a small sample of the pulverized ore in a test-tube, add to it some nitric acid, a little more than will cover it, and heat till the acid does not seem to dissolve any more of the ore ; let cool, after which add as much *pure water* as there is acid, and shake.

Filter, in manner described under "Copper Analysis," p. 349.

Sulphates.—To some of the filtered acid solution add solution of chloride of barium (or, if lead be present, of nitrate of barium). A white cloudiness or precipitate (which does not instantly form in dilute solutions) shows the presence of sulphates.

Sulphides.—To a piece of the rock, or to some of the powdered ore, add a drop of nitric acid. If sulphides are present in any quantity, a strong odor, similar to that of rotten eggs, will be given off.

Tellurides.—Take a small piece of the ore and place it on the cover of a porcelain capsule, and heat with the inner flame of the blow-pipe for a couple of minutes. Now place a drop of *concentrated* sulphuric acid on the cover, and let it slide down to the heated fragment. As soon as it touches or approaches very near the ore a beautiful carmine coloration forms, strongly contrasting with the white porcelain. As the latter cools, the color fades. Any white crockery, as a piece of a broken plate or saucer, will do to use in this test.

Copper.—To a piece of the rock on a white porcelain surface add a few drops of nitric acid and stir. Add now an excess of ammonia water. If the mass turns blue, copper or its compounds is undoubtedly in the ore. If the latter contains much copper, a polished knife-blade dipped in an acid solution of it will receive a coating of metallic copper.

Iron.—If, at the same time the solution treated with ammonia turns blue, or even if it does not do so, there appears on the porcelain or in the test-tube, a reddish-brown gelatinous mass, then iron is present.

As further tests for iron, on one part of an old plate put a crystal of sulphocyanide of potassium and on another a lump of ferrocyanide of potassium (yellow prussiate of potash);

now pour on each a little of a hydrochloric acid solution of an ore containing iron ; a blood-red coloration with the first-named re-agent, and a magnificent blue precipitate with the second, prove conclusively the presence of iron compounds. Of course these tests can be shown with the filtered solution in test-tubes.

Lead.—Drop a little nitric acid upon a piece of ore supposed to contain lead, then add a little water, and finally a crystal of iodide of potassium. A bright-yellow precipitate will form if lead is present.

Manganese.—The best and simplest test for manganese is to fuse the substance with a little carbonate of soda and nitrate of potash on a strip of platinum in a hot flame. The manganese unites with the sodium, forming *green* manganate of soda. The manganese may have to be separated from other matters in a manner similar to the method given under the crucible process, p. 255.

Silver.— If this metal is in any appreciable quantity in an ore, it will dissolve in nitric acid (excepting the chloride ores). To the acid solution add a little hydrochloric acid, solution of common salt, or even a dry grain or two of the latter. A curdy, white precipitate of chloride of silver is thrown down, which is not soluble in water (as is chloride of lead, on the contrary), but dissolves easily in ammonia water. The precipitate turns black on being exposed to light.

As stated above, chloride ores do not dissolve in nitric acid ; therefore, when they are suspected to be present, put some of the powdered ore into a small bottle, pour in a small quantity of very stong ammonia water, cork up the bottle, and let it stand for a few hours. Then add, in slight excess, nitric acid. The white precipitate of silver chloride will at once come down if there is any in the ore.

The best test for gold is the fire assay. To learn the colors and appearances of the tests above given, try them on the following substances :

Carbonates...............Bi-carbonate of Soda.

Sulphates.....................Sulphuric Acid.

Sulphides...............Copper or Iron Pyrites.

Tellurides ....Any Telluride Ore.

Copper ....Copper Wire.

Iron ....Nail or Wire.

Lead ....Sheet Lead or Galena.

Silver ....Silver Foil and Horn Silver.

Manganese..........Black Oxide of Manganese.

Consult the books on qualitative analysis for further information or tests.

XI. BRIEF SCHEME FOR SILICA, IRON, AND MANGANESE.

It is very often the case that the percentages in an ore of the above-mentioned substances are wanted. More particularly is this true with carbonate ores. Hence the following notes :

Dissolve the weighed ore in hydrochloric acid by the aid of heat. Filter hot, and wash with hot water. The filtrate contains the iron, with chloride of lead, etc.

The silica on filter contains chloride of lead.

Wash this out with hot solution of citrate of ammonium, following with hot water. Ignite the silica while still damp.

To the iron in solution in the filtrate add sufficient sulphuric acid to convert all the lead into sulphate of lead. Warm the solution, if not already so, and add, drop by drop, dilute stannous chloride solution, until the liquid becomes colorless, showing that the iron is all reduced to state of protoxide. Avoid a great excess of the tin solution. Now cool, and add, *all at once*, an excess of strong mercuric chloride solution. The precipitate formed should be perfectly white. If dark-colored, it indicates that insufficient mercuric chloride has been used, and the analysis is spoiled. If the precipitate is all right, the solution is ready for titration with standard bi-chromate of potash solution. (Consult Fresenius' "Qualitative and Quantitative Analysis," and Hart and Sutton on "Volumetric Analysis.")

For manganese in ores (excepting silicates), heat a weighed sample in crucible in open fire for fifteen minutes, converting the manganese into protosesquioxide of manganese. Treat with hydrochloric acid, and titrate with iodide of potassium and hyposulphite of sodium. (See Sutton.)

XII. DETERMINATION OF MOISTURE IN AN ORE.

It is often a matter of importance to know the amount of moisture or water contained in an ore. The simplest manner in which to determine this, and a satisfactory one at that, is to sample out a certain weight, say *five grammes*, and transfer to a porcelain capsule, the weight of which is already known. Expose the capsule and contents to steam heat, in any convenient way, for one-half hour, then weigh. Heat half an hour longer and weigh again. There should be but a slight difference in the last two weighings. The difference between the last weight and the original weight of dish and ore is the loss by driving off the water; this difference divided by the amount of ore taken, and multiplied by 100, is the percentage of moisture in the ore.

XIII. DETERMINATION OF SULPHUR IN PYRITES.

Weigh 1 gramme (or say 10 grains) of the finely powdered ore into a casserole; add a small amount of chlorate of potash, cover with watch glass, add 50 c.c. concentrated nitric acid, and heat to boiling, adding a little more chlorate from time to time. When perfectly oxidized, remove watch glass (and it should be rinsed into the casserole), and evaporate to small bulk on a water bath. Add a little strong hydrochloric acid, and evaporate to dryness, moisten with the same acid, add water and filter from silica and the gangue.

To the filtrate add 1 gramme (or 10 to 15 grains) tartaric acid, heat, add hot solution of baric chloride, drop by drop, boil, let settle, filter and wash well with hot water.

Weigh a clean porcelain or platinum crucible, add filter and precipitate, burn to ashes, cool, weigh as baric sulphate: after deducting weights of crucible and filter ash, multiply remainder by 0.1374, and the product by 100 or 10 for percentage of sulphur.

XIV. THE ASSAY OF TIN ORES.

In order to make an accurate fire-assay of a tin ore, it is necessary to bear in mind several things. First, that the cassiterite or tin oxide is itself a small percentage of the ore ; secondly, that the other matters present tend to prevent an accurate result; it is therefore essential that the tin ore should be previously concentrated and purified.

Preparation of the Sample.—Weigh say 50 lbs. of the ore, and remember that the resulting "concentrate" is to be also weighed, so that we may know its proportion to the original ore. Crush, by means of hand crusher, motar and rubbing plate until the whole, save the mica, will pass through a 20-mesh sieve. Examine the mica for any attached mineral, remove by grinding and screening and add to the first lot screened. The mica is to be rejected. "Pan" the screenings, being careful that the tailings do not contain any tin.

The above treatment gets rid of about all

impurities save garnets, sulphides and selenides. Roast the concentrates in a roasting dish as usual, and after roasting, chill in cold water. Boil the roasted ore in a porcelain dish with nitro-hydrochloric acid (removing iron), wash off acid and any light particles, dry and weigh. If, for example, the 50 lbs. of ore should produce ½ pound of finally purified ore, it would represent 1 per cent. of the original ore. The next step is to ascertain how much metallic tin the supposed tin oxide contains.

Make up the following

CHARGE.

Ore......................... 10 grammes or 160 grains.
Cyanide of potash.......... 40 　 " 　 " 640 grains.

Use size " J " crucible, ram 5 grms. (80 grains) cyanide of potash in bottom, then the charge, lastly 5 grms. cyanide as a cover, *us ng no salt.* Time, 15 minutes.

The fire should be *hot* and "kept at the highest point to which the cyanide can be heated without beginning to boil or to evolve heavy fumes. An assay may be considered finished when the pure upper slag has become so transparent that the impurities contained at the bottom of the crucible are visible through it. The tin will then have collected into one button beneath the lower slag, and very few, if any, frills will be found."—(*Hoffmann.*)

The assay of a tin ore needs much care and practice, and the cyanide should be the *best* grade (98 per cent.) for, although this is expensive, its use will give the most accurate results.

Assayers and others especially interested in tin ores will find a most exhaustive and interesting letter on " The Dry Assay of Tin Ores," by Prof. H. O. Hoffmann, which was read at the Colorado meeting of the American Institute of Mining Engineers, June, 1889, and given in the Technology Quarterly, Vol. 3, No. 2, and in the Chemical News, in the Nos. ranging from Aug. 1, to Dec. 26, 1890, from which article this chapter is chiefly extracted.

XV. GOLD AND SILVER ORES AND MINERALS.

   I. How to *find* them.

   II. How to *know* them.

   III. How to *value* them.

   IV. How to *treat* them.

I. How to *find* them.

A careful study of the entire subject of gold and silver ores and minerals should start at the very beginning. This means then, first, a little knowledge of Geology in general, and, to learn *how* these ores and minerals came to be deposited in their several places, necessitates the study of Geology in particular, namely: the formation of ore deposits. The would-be prospector can do no better with his spare winter evenings than to occupy them in reading the authorities on page 474, in perhaps the following order: Dana (text-book), Rutley, and Von Cotta, in full, and referring to Dana (manual), Le Conte, and to such other kindred books as the judgment selects, for special information on ore deposits.

The above for the winter. In the spring

the prospector wants to get into the mountains. A useful little work to read just before starting is Pomeroy's Mining Manual for Prospectors and Miners (page 477 this book). Note especially his "Camp Outfit for Three," on page 69.

Arriving at his destination, the prospector begins his search. The following article on "How to Prospect," copied from Wilson's Mining Laws (originally in Blake's Handbook of Colorado) will be good, seasonable advice, and read in connection with what Pomeroy says on "Prospecting," page 58, will give him considerable aid:

1. Examine the gravel and boulders of the mountain streams, and note carefully the structure and character of the gravel wash. This will reveal the geological formations that are intersected by the stream. Try the sands at the head of the gravel bars for free gold,* or for any crystallized mineral If the structure of the quartz boulders or other vein stones is favorable, go up the stream until the geological zone is found that has produced the quartz or other metal-bearing minerals. Then follow the supposed metal-bearing zone on its

---

* See " Pan Test for Gold" (" Panning "), on page 369.—(W. L. B.)

line of strike,*' and make especially careful examinations wherever eruptive dykes are found intersecting the formation.

2. When a lode or vein is found, note carefully its relation to the country rock, especially any differences in the opposite walls of the vein. Then follow it on the line of outcrop and note carefully those points where the best † ores are seen, so as to determine the position of the best ore chutes before making any location on the lode.

3. The first work should consist of shallow cuts across the lode at intervals of 50 to 100 feet, or if the vein is small and partially covered by soil and debris, a trench along the line of out-crop is preferable. If the surface tracing is satisfactory, and the true line of strike has been determined, then survey your claim and stake off the boundaries according to the requirements of the United States laws. ‡

4. The work of exploring the vein under ground is the next thing in order. To do this intelligently you must select that point on the line of outcrop where the best ore is found, then sink a shaft on the lode, following the angle of dip, keeping both foot-wall and hang-

---

* Glossaries of terms used in geology, mineralogy, prospecting, mining, metallurgy, etc., will be found in Pomeroy's Mining Manual (pp. 72-9) Copp's Mining Code (pp. 88-125, by R. W. Raymond), Collins' First Book in Mining and Quarrying (pp. 101-114), Phillips' Explorers' and Assayers' Companion (pp. 423-468), Dana's Manual of Mineralogy (pp. vii-xii), and Anderson's Prospector's Handbook.—(W. L. B.)

† Confirming such observations by assays.—(W. L. B.)

‡ Pomeroy (pp. 80-116), also note list of books on Mining Law, in appendix (pp. 435-6).—(W. L. B.)

ing-wall exposed if possible. If the lode is too wide for this to be done, then follow the best ore streak of the vein itself, and at every fifty feet in depth make cross-cuts to the walls of the vein.

5. After a 100 feet deep shaft has been reached, run levels each way from the shaft on the line of the vein in order to determine the extent or spread of the ore chute or chimney on the horizontal line. When the limit of the ore body on the horizontal line has been ascertained, then sink 100 feet more and drift right and left as before. If more than one chimney of ore is found on the line of the vein, a shaft should be sunk on it, and drifts run as above stated, being careful to confine all the exploring work within the walls of the vein itself. *

6. When enough has been done to prove the character, size and quality of the vein, it will then be time to determine the position, character and extent of the " dead-work " necessary to work the mine to the deep.

---

*Concerning the "sampling" of a vein or the side walls of a tunnel or indeed of any part of a mine, which may be well to have done frequently, proceed as follows: On a distance, say of 300 feet, divide the exposure into three sections of 100 feet each. Then taking a certain height from floor of tunnel. conveniently breast-high, chip *continuously* pieces in a vertical width of from 3 to 6 inches, letting the pieces drop onto paper or bagging spread on the floor of tunnel at bottom of side wall, and so go along the entire 100 feet. The three or four hundred pounds thus obtained, will represent a fair average of all veins vertically crossing the wall, and are to be crushed and averaged down as usual. And the same with the remaining two sections of 100 feet each. By such sampling, the miner will be sure to get at least portions of any rich veins or masses exposed. Should one section show up especially well, divide that section in halves, and sample again. By such a series of eliminations, a rich " find " may be discovered, where a mere hand-picked specimen might miss it altogether. Above all, beware of the " *salting*" of a mine, by interested persons. A word to the wise is sufficient.—(W. L. B.)

These questions should be settled by careful surveys made in the light of all the local facts and surroundings, such as the geological structure of the country rock, the probable amount of water to be raised, the lowest point of drainage by adit or level, and the most convenient point of delivery of the ores to the surface, etc.

The last part of the preliminary exploration of any mine is to determine, by actual tests, what are the best methods of reduction,* and the extent and kind of reduction works needed, etc.

7. After all the preliminary facts have been thoroughly ascertained and clearly defined, the unavoidable risks of mining will have been fully met and overcome. All subsequent operations are simply matters of skill , and business management, and the capitalizing of the mine becomes a mere matter of business detail.

The requirements are as follows:

1. The preliminary exploration must have ore enough cut and under-run, or otherwise exposed, to give at least two years' work for reduction work of an extent sufficient for the annual average out-put of ore.

2. The reduction works must be suited for the best treatment of the ore.

3. The exploration of the mine must be pushed ahead of the extraction of ore, so as to expose at least one ton of ore in new ground for every ton extracted from the previously exposed ground.

4. Before erecting reduction works, the exposed ore

---

* See Section 4, " How to Treat Them."—(W. L. B.)

in the mine should be so thoroughly tested as to guarantee a net profit to pay the whole cost of such work.

5. The mine being well opened, and the reduction works, or plant, established, the general success of the enterprise must depend upon the efficiency of the general business management.

A little pamphlet by F. L. Bartlett, entitled "Minerals of New England, Where and How to Find Them," while somewhat local, is not enough so to prevent it giving fair descriptions of the minerals found elsewhere, with tests for the same. See also, "Underground Treasures: How and Where to Find Them," by Jas. Orton, and "The Prospectors' Handbook," by J. W. Anderson.

II. How to *know* them.

That science which covers the knowledge of *rocks* is termed *Lithology*; that which concerns *minerals* is *mineralogy;* while *crystallography* restricts itself to the specific structure of rocks and minerals. The study of the above three branches of exact science gives the theoretical knowledge which is to be transmitted into practical results by aid of actual work in blow-pipe analysis and determinative mineralogy, combined with qualitative and quantitative analysis.

All this seems formidable enough, but no one can become a thorough expert, unless he does understand well the above named branches, hence it means patient study, long practice and a certain amount of manipulative skill. But even a smattering of all of the "ologies" outlined, will not injure the miner or prospector, and is sure to help him. There is no royal road to learning these sciences, but attention, observation, experience and experimenting will ultimately enable him to learn very much about the nature of ores and minerals.

Gold and silver, one or both, occur free or in combination, or associated with many rocks, ores and minerals. The following list gives a slight idea of the universal distribution of the precious metals in and among other minerals, ores and rocks:*

---

*See lists pp. 181-4, also pp. 461-5 in Appendix

Aikinite, algodonite, altaite, anatase, apatite, argillaceous schist, arsenopyrite, asbestos, barnhardite, barytes, beresite, beryl, bismuth-inite, bismuth ores, blende, bornite, boulanger-ite, brookite, calcite, cerussite, chalcopyrite, chlorite, chloritic - schist, chloro-arseniate of lead, chrome-iron, chrysocolla, copper ores generally, copper pyrites, corundum, cuprite, diamond, diorite, emerald, enargite, feldspar, ferro-tellurite, fluor-spar, galena, garnet, gneiss, granite, heavy spar, hematite, horn-blende, horn-blendic schist, iridosmine, iron oxides, iron pyrites, itacolumite, joseite, kyanite, leucopy-rite, magnesite, magnetite, malachite, man-ganese oxides, melaconite, mica slates, mime-tite, mispickel, molybdenite, monazite, native antimony, native arsenic, native bismuth, na-tive copper, native mercury, native palladium, native platinum, native tellurium, porphyry, pyrite, pyrites of various kinds and mixtures, pyrrhotine, quartz, realgar, red and brown hematites, red oxide of copper, rhodium, ruby, rutile, sapphire, scheelite, selenpaladite, serpentine, siderite, smaltite, sphalerite, spinel, steatite, stibnite, talc, talcose-schist, tellurite, tellurium ores, tellurpyrite, tenorite, tetrady- mite, tetrahedrite, tinstone, titaniferous iron, topaz, tourmaline, trap rock, wehrlite, zinc- blende, zircons.

But the miner holds in his hand a piece of ore and wants to know the probabilities. What are they? Aside from actual tests, all that can be given are hints and suggestions, indicat- ing the probabilities. If the sample is *very* heavy, look out for lead (it might also be an ore of mercury, or common barytes), and if also it is yellowish, it is pretty sure to contain lead. If moderately heavy and is yellowish, brownish or reddish in color, there is undoubt- edly iron present. If very black, probably manganese. If blue or green, suspect copper (a very light green, certain arsenic ores). The above describes the specimen when it is of a *dull, earthy* appearance. If, however, it is shiny and light yellow (brassy), it will be or contain iron pyrites; if of a more golden yellow, or of a rainbow shimmer, then copper pyrites; if leaden, there is galena, if *whitish* lead color, blende. Examine ores and minerals known to be those named above, and remember their characteristics. And further, try to confirm your suspicions by the simple tests given on pp. 422-427 in the appendix. Metallic silver and gold are too well known, even to the ignorant, to be mistaken for other metals, but by crushing and panning, or by fusing in a crucible according to processes described in the chapter on assaying, all doubt will be removed.

And this is about all that one can say in trying to teach the nature of a gold and silver ore by simple observation only, without the corroboration of the blowpipe, the test-tube or the crucible.

III. How to *value* them.

The most common, if not the very first question put to the assayer (and one which he is expected to answer at once), is: " What will it carry?" referring to the gold and silver value of an ore or mineral,　A very easy question to ask — an exceedingly difficult one to answer.

Of course the only *accurate* answer is that given by a careful assay, but what can we say without first using this infallible test?　The specimen may belong in one of two divisions; it may be a typical gold or silver mineral, or an ore of a class known to often carry the precious metals.

Thus, of the first class, the sample under discussion may be free gold in some one of its various forms, native silver, ruby silver, horn silver, silver-glance, etc.　Now the mineralogist knows that if the mineral is pure silver-glance (with no gangue) it must contain 87 per cent. silver—over 25,000 ounces to the ton, (provided a ton of it were found); if it is pure horn-silver, which is 75 per cent. metal, then a ton of it would increase the wealth of the owner by nearly 22,000 ounces; the ruby silver, according to whether it was the dark or the light-red, the antimonial or the arsenical variety, would pan out in round numbers, from 17,000 to 19,000 ounces per ton.

If the sample is the pure mineral in a simple gangue, then the specific gravity will give the value. It is thus that free gold in quartz is sometimes estimated.

If the sample is a mixture of various gangues and minerals, but still shows that it contains one or more of the valuable minerals mentioned, then no exact figures of value can be given without an assay, but beforehand the comforting assurance can be truthfully put forth, that the ore is *rich*.

But when we come to the second class, of ores proper, we are obliged to proceed very carefully. Among the most common sources of gold are the sulphurets of copper and iron, being iron sulphuret alone (iron pyrites) or copper and iron sulphuret (copper pyrites or chalcopyrite). We call either, when gold-bearing, *auriferous sulphurets*. The gold in these pyrites may vary in quantity from an infinitesimal trace, a practical nothing, up to many ounces per ton, and yet there be nothing on the surface to indicate the degree of richness. There is very little use in guessing the value of pyrites from its appearance only.

The *locality* is a good indicator, but that often fails. If the pyrites was found in one of the gold-mining districts of the West, the South or even the extreme East, it may be valuable — it may likewise be worthless. But if found in other parts of the country, the chances are that the gold exists in too small quantity to be worked with profit.

It is dangerous to generalize from specific things, but I can not resist the temptation of saying, as drawn from my experience, with regard to these sulphurets, that nine out of every ten will be *low in gold*.

In short, with any gold ores, unless the gold actually shows itself, beware of guessing at their value.

With ores carrying silver, we have a little better chance. Galena, blende and "carbonate ores," carry more or less silver. Concerning the last class, surface indications are worth but little. They resemble so many valueless earths, that an assay, only, can positively determine their value. Blende (or zinc sulphuret or "black jack") seldom bears much silver. And more, I have found that ores that contain much blende, no matter what other minerals are present, do not often run rich in silver.

Galena is rather a perplexing mineral, so far as it shows or does not show silver. The old theory that a fine-grain or "steel" galena was invariably rich, and a coarse-grain poor, is now pretty well exploded. The fine-grain is as often poor as rich, and the coarse-grain, instead of being low in silver, may mount up into the hundreds. Study a piece of galena, and see if it possesses any one of the three following characteristics:

1st. Does it have a peculiar bluish-purple lustre?

This lustre, difficult to describe in words, is

easily perceived, and is entirely different from the ordinary leaden shimmer of pure galena. The ores possessing it are pretty sure to be rich.

2d. Does it have a smooth, almost curving surface, an appearance not unlike plumbago, and yet, at the same time, appears as if one could run his knife-blade just under the surface, and raise up a scale, as of mica, but yet one cannot?

Such ores are invariably rich.

3rd. Does it show, perhaps in the cleavage, a greenish coloration, more or less distinct? It will be rich. It does not follow that the silver is in combination with the galena; it may be a distinct mineral simply *in* the galena. But this is immaterial. The galena, as an ore, is rich, and that is the vital point.

Furthermore, galena in a siliceous gangue is much more likely to be rich than that which is in a lime-stone (calcite) gangue.

Gray copper (tetrahedrite) is a convenient name, applied, ordinarily, to any gray mixture of sulphurets. Rich or poor, in either gold or silver, it puts outside no sign of its varying possibilities.

Here, then, we have opinions based largely upon experience, upon the characteristics of minerals which have shown by assay tests that they are what these indications have pointed out.

IV. How to *treat* them.

If the question which headed the last section, viz.: "What will it carry?" is the first one put to an assayer, the second one will surely be, "How would you treat this ore?" And this is even more difficult to answer at sight than was the first.

Before undertaking to treat an ore, we should know four things:

A. The assay value of an average of the ore.

B. The nature of the metalliferous mineral or minerals containing the silver or gold, or both.

C. The nature of the gangue.

D. The relative proportions of the gangue and the metal-bearing minerals.

A. The determination of the assay value of an *average* of the ore, or a mill-run checked by an assay will settle the question as to whether the ore is *high grade* or *low grade.*

B. The nature of the metalliferous mineral or minerals containing the silver or gold or both.

The precious metals most commonly occur in some one or more of the following groups:

I.
Free Milling Ores.

1. Free gold and free silver.
2. Chlorides, bromides, iodides or mixtures of them.
3. Silver glance or sulphuret of silver.

II.
Smelting Ores.

{
4. Oxides, carbonates, or mixtures of both, of lead, iron or copper.

5. Sulphurets of iron, copper, or iron and copper or other mixed sulphurets (as gray copper). Also galena and blende.

6. Arsenical and antimonial ores (tellurides, selenides, etc.).
}

There are other and rarer minerals not included in the above, but the classes are reasonably complete. The classification is arbitrary, but the limits are not always sharp and clear, consequently it is not necessarily implied that any given ore must belong to but one of the divisions named. On the contrary, it may be of such complex nature as to belong to two or more, or even all six of the classes, consequently be both free milling and smelting, and whether either free milling or smelting, it may contain representatives of the three classes included under the subdivision in question.

C. The nature of the gangue.   (See pp. 239-242.)

D. The relative proportions of the gangue and the metal-bearing minerals.

General rule: " The heavier an ore, the greater the percentage of metalliferous mineral; the lighter an ore, the greater the percentage of gangue."   There are a few exceptions to this rule.   The chief one commonly met with is barytes or heavy spar, a gangue which is three-fifths as heavy as galena, the chief ore of lead.   (Pp. 238-239).

The relative proportions of gangue and mineral determine whether or not an ore is to be concentrated.   The eye tells something of these proportions, but a " panning " test will give quantitative results.

The conditions of outlook have now been mentioned, uninfluenced by the questions of locality of the mine producing the ore, transportation, fuel, water, price of labor, etc. These, although very important, or indeed deciding whether or not an ore can be treated on the ground, will come in place later on.

An inexperienced mining man may now say that it is all very well to make the distinctions drawn, but how is he to know the value of the ore, or the particular form of combination of the gold and silver, or the kind of gangue, or whether it is necessary to concentrate, etc. The only truthful answer that can be made is to assure him that in the long run it will undoubtedly pay him to secure the services of those trained in the professions of mining engineering, metallurgy, assaying and chemistry, and who are devoting their time, brains and energies to such work.

But either with or without such aid, there is no harm in his learning all he can about ores, and we will start out with an example, as being the simplest, most direct, and convincing of arguments.

Let him behold, then, quite a little lot of ore, representing a fair average of the product of the mine from whence it came, some of the pieces coming from just below the surface, others from varying depths in the vein. They also represent the width of the fissure, for certain of the pieces are from next the hanging wall, certain from close by the foot-wall, with others from the centre of the vein.

How shall we ascertain the nature of this ore? *Long experience* in the visual examination of ores and minerals, blow-pipe tests, qualitative analyses, and assays will tell this. It is not to be expected that the ordinary mining man, or man of business who goes into mining, will know much about the latter branches of investigation (although even a slight acquaintance will benefit him). But as to the other qualification, the very word "experience" will show him how he will learn; by being told what any particular mineral is, and what any certain ore looks like, by a good memory, and a quick and searching eye; by good use of his observant faculties, he will in time accumulate a vast amount of information which he can wisely supplement with a little study of the authorities.

To return to the ore. This time it happens to be very simple in composition, being merely a quartz rock, deeply stained with reddish brown oxide of iron. Breaking open several lumps, and examining the interiors with a magnifying glass, shows no signs of sulphurets. These may have been, and probably were once in it, but not at the present time—they have all been oxidized. Besides the iron oxides, it is probable that there is also a little carbonate of iron present, but that makes no difference. Further, it is possible that magnetic oxide of iron is present. Such an ore is typically gold-bearing. It may be rich (high grade) and show the gold. It may be rich and show none of the yellow metal. It may be poor (low grade). It may not contain even a trace of the precious metal. Here the eye ceases to be the arbitrator, and re-

course *must* be had to something more defi-
nite in its verdict, and that is an assay.

Moreover, if gold *is* present, the chances are
that more or less silver will be associated
with it.

The average result of several careful assays
gives an assumed mean of $\frac{55}{100}$ oz. of gold,
and ¾ oz. of silver. With gold at $20.67,
and silver at $1.29 the ounce, this ore would
be worth $12.33 per ton. The actual market
value of the gold and silver would be even less
than this, say $18 to $20 per oz. of gold, and
silver according to market quotations on the
day of sale. But, assuming the ore to be
worth, as stated, $12.33, local circumstances
will decide whether it is high grade or low
grade.

We have then ascertained the average
amounts of gold and silver that are contained
in the ore (and the hypothetical value which
is to be calculated down to the market value),
the nature of the mineral (*i. e.* free gold), and
the nature of the gangue (*i. e.* iron-stained

quartz). This is all we care about at present.

The next point is the manner of treatment. The ore is essentially of the first class, that is, free milling, for there are in it no refractory minerals, as blende, pyrites, etc.

The small amount of silver would seem to indicate that it is not present as a sulphuret or as some one of the chloride group (which fact could be settled by a careful analysis), hence the legitimate deduction is that both the precious metals exist *at present* in the free state, whatever they may have been previously. The metallurgical treatment then to be adopted is that known as free milling, or direct amalgamation. Assays of the ore before amalgamation, and of the tailings will determine the percentage of extraction of the gold and silver by the particular form of milling machinery adopted. This percentage will serve as a check (it will be always higher) on the actual product obtained.

But suppose that by visual examination or

by blow-pipe we learn that the sample contains *sulphurets*, as well as the iron oxides (and indeed it is very usual to find sulphurets appearing in ores as the work progresses in or down, for the oxidation may be only at the surface), then what? Tests must show how much gold is free-milling, and how much passes away with the tailings (that is in and among the sulphurets). If the amount found, by frequent average tests, in the sulphurets will pay to work (and local conditions only can determine this), then smelting, preceded by concentration, must come in. The following table is from an actual "milling test:"

	GOLD, OZ. PER TON.	GOLD, VALUE PER TON.	SILVER, OZ. PER TON.	SILVER, VALUE PER TON.	TOTAL VALUE PER TON.
Assay of the Ore.	$\frac{6}{10}$	$16.54	$\frac{2}{10}$	$0.20	$16.74
Saved by Free Amalgamation.	$\frac{882}{1000}$	$14.25	$\frac{2}{10}$	$0.20	$14.45
Saved, per Ton of Original Ore, by concentration to $\frac{3}{10}$ of 1 per cent.	$\frac{13}{1000}$	$ 0.24	Trace.	—	$ 0.24
Passed into Tailings.	$\frac{99}{1000}$	$ 2.05	Trace.	—	$ 2.05
Assay of Sulphurets.	$4\frac{2}{10}$	$86.82	$\frac{2}{10}$	$0.20	$87.02

In general, such an ore as the above would pay to work.

This is not a treatise on smelting, and we can give but a little advice: Get a man who understands his business thoroughly, to manage your property, if it proves to be worth working, and if the ores are complex. And what is here written with regard to iron oxides and sulphurets will be equally true of any other ores which are both free-milling and refractory.

XVI. DETERMINATION OF SPECIFIC GRAVITY OF MINERALS.

1st. *By the balance.*—Weigh a small piece of the mineral on a delicate balance, next place a little bench straddling the scale pan (without touching it) and on this set a small beaker of distilled water of a temperature of 60° F. Tie a piece of hair around the mineral and suspend it in the water from the scale-beam hook, so that mineral does not touch sides or bottom of beaker, shaking off all air bubbles and weigh again. Formula as follows:

Weight of mineral in air $= W$

" " " " water $= W'$

$$\text{Specific gravity} = \frac{W}{W-W'}$$

2d. *By the flask.*—Take a specific gravity flask, as shown in Fig. 131, and fill it with distilled water of 60° F., so that when the stopper is squeezed in, the water will over-run through the narrow opening at top, and weigh carefully. Powder some of the mineral, empty a little of the water from the flask, pour in the powdered mineral, fill with water, tighten stopper as before, remove excess of water, and again weigh. In both weighings the bottle is "Tared" by an equal weight on the other scale-pan.

FIG. 131.

Formula:

Weight of mineral $= W$

" " flask and water $= W'$

" " flask, mineral and water $= W''$

$$\text{Specific gravity} = \frac{W}{(W + W') - W''}$$

SECTION II.

LISTS AND REFERENCES.

GOLD.

LIST OF THE PRINCIPAL GOLD MINERALS FOUND IN THE UNITED STATES.

NAME.	COMPOSITION.
1. Calaverite (telluride of gold).	Gold, tellurium.
2. Gold amalgam.	Gold, mercury.
3. Electrum (argentiferous gold).	Gold, silver.
4. Müllerite (telluride of gold, silver and lead).	Gold, silver, lead, tellurium.
5. Nagyagite (black tellurium, foliated tellurium, telluride of gold and lead).	Gold, lead, tellurium (antimony, sulphur).
6. Native gold (flour, leaf, wire, nugget, free, etc.).	Gold.
7. Petzite (telluride of gold and silver).	Gold, silver, tellurium.
8. Sylvanite (graphic tellurium, yellow tellurium, telluride of gold and silver)	Gold, silver, tellurium (antimony).

MINERALS LIKELY TO CARRY GOLD.

1. Aikinite.	14. Magnolite.
2. Altaite.	15. Melaconite.
3. Argentite.	16. Native arsenic.
4. Arsenopyrite.	17. " bismuth.
5. Bismuthinite.	18. " silver.
6. Chalcopyrite.	19. " tellurium.
7. Coloradoite.	20. Pyrite.
8. Ferro-tellurite	21. Sphalerite.
9. Galenite.	22. Tellurite.
10. Henryite.	23. Tellurpyrite.
11. Hessite.	24. Tetradymite.
12. Joseite.	25. Tetrahedrite.
13. Lionite.	26. Wehrlite.

SILVER.

LIST OF THE PRINCIPAL SILVER MINERALS FOUND IN THE UNITED STATES.

NAME.	COMPOSITION.
1. Alaskaite (sulphide of bismuth, silver and lead).	Silver, bismuth, lead, copper, sulphur.
2. Argentite (sulphuret or sulphide of silver, vitreous silver, silver glance).	Silver, sulphur.
3. Bromyrite (bromide of silver, bromic silver).	Silver, bromine.

4. Cerargyrite (muriate or chloride of silver, horn-silver) — Silver, chlorine.

5. Dyscrasite (antimonial silver). — Silver, antimony.

6. Electrum (argentiferous gold). — Silver, gold.

7. Embolite (chloro-bromide of silver). — Silver, chlorine, bromine.

8. Freieslebenite (antimonial sulphide of silver and lead). — Silver, lead, antimony, sulphur.

9. Hessite (telluride of silver, telluric silver). — Silver, tellurium.

10. Iodyrite (iodide of silver, iodic silver). — Silver, iodine.

11. Miargyrite (sulphide or sulphuret of silver and antimony) — Silver, antimony, sulphur.

12. Native silver (free, wire, leaf, dendritic, etc.). — Silver.

13. Petzite (telluride of silver and gold). — Silver, gold, tellurium.

14. Polybasite (sulphide of silver, antimony and arsenic). — Silver, antimony, arsenic, copper, sulphur.

15. Proustite (arsenical silver ore, light red silver ore, ruby silver). — Silver, arsenic, sulphur.

16. Pyrargyrite (antimonial red silver ore, dark red silver ore, ruby silver). — Silver, antimony, sulphur.

17. Schapbachite (bismuth-silver, sulphide of bismuth, silver and lead). — Silver, bismuth, lead, sulphur.

18. Schirmerite (same as above but proportions varying). — Silver, bismuth, lead, sulphur.

19. Stephanite (sulphide of silver and antimony, brittle silver, black silver). — Silver, antimony, sulphur.

20. Sternbergite (sulphide of silver and iron). — Silver, iron, sulphur.

21. Stetefeldite (oxide of antimony with silver, etc.). — Silver, antimony, copper, oxygen, sulphur.

22. Stromeyerite (sulphide or sulphuret of silver and copper, silver-copper glance). — Silver, copper, sulphur.

23. Sylvanite (graphic tellurium, yellow tellurium, telluride of silver and gold). — Silver, gold, tellurium (antimony).

24. Tetrahedrite (gray copper ore, sulphide of copper, antimony, silver, etc.). — Silver, copper, antimony, sulphur (arsenic, bismuth, mercury, zinc, etc.)

MINERALS LIKELY TO CARRY SILVER.

1.	Algodonite.	22.	Melaconite.
2.	Altaite.	23.	Müllerite.
3.	Arsenopyrite.	24.	Nagyagite.
4.	Barnhardite.	25.	Native antimony.
5.	Bornite.	26	" arsenic.
6.	Boulangerite.	27.	" bismuth.
7.	Calaverite.	28.	" copper.
8.	Cerussite.	29.	" gold.
9.	Chalcopyrite.	30.	" mercury.
10.	Coloradoite.	31.	" tellurium.
11.	Enargite.	32.	Petzite.
12.	Ferro-tellurite.	33.	Pyrite.
13.	Galenite.	34.	Realgar.
14.	Geocronite.	35.	Smaltite.
15.	Gold amalgam.	36.	Sphalerite.
16.	Henryite.	37.	Sylvanite.
17.	Hessite.	38.	Tellurite.
18.	Joseite.	39.	Tellurpyrite.
19.	Leucopyrites.	40.	Tetradymite.
20.	Lionite.	41.	Wehrlite.
21.	Magnolite.		

COPPER.

LIST OF THE PRINCIPAL COPPER MINERALS FOUND IN
THE UNITED STATES.

NAME.	COMPOSITION.
1. Aikinite (needle ore, acicular bismuth, cupreous bismuth).	Copper, bismuth, lead, sulphur.
2. Algodonite (arsenide of copper).	Copper, arsenic.
3. Atacamite (muriate of copper, oxy-chloride of copper).	Copper, chlorine, oxygen (water).
4. Aurichalcite (carbonate of zinc and copper).	Copper, zinc, carbon, oxygen (water).
5. Azurite (mountain blue, blue carbonate of copper, blue malachite, azure copper ore).	Copper, carbon, oxygen (water).
6. Barnhardite (sulphide of iron and copper).	Copper, iron, sulphur.
7. Bornite (purple copper ore, variegated copper ore, *erubescite*, sulphide of copper and iron, horseflesh ore).	Copper, iron, sulphur.
8. Bournonite (triple sulphuret of copper, lead and antimony).	Copper, lead, antimony, sulphur.
9. Brochantite (sulphate of copper).	Copper, oxygen, sulphur (water).

10. Caledonite (cupreous sul-phato-carbonate of lead).

Copper, lead, car-bon, oxygen, sul-phur.

11. Carrollite (sulphide of cobalt (nickel) and copper).

Copper, cobalt (nickel), sulphur.

12. Chalcanthite (blue vitriol, copper vitriol, sulphate of copper).

Copper, oxygen, sulphur (water).

13. Chalcocite (copper glance, vitreous copper, sulphuret or sulphide of copper).

Copper, sulphur.

14. Chalcopyrite (copper pyrites, pyritous copper, sulphide of copper and iron).

Copper, iron, sul-phur.

15. Chrysocolla(mountain green, mountain blue, silicate of cop-per).

Copper, silicon, ox-ygen (water).

16. Covellite (indigo copper, blue copper, sulphide of copper).

Copper, sulphur.

17. Cuprite (red oxide of copper, cupreous oxide, tile ore).

Copper, oxygen.

18. Domeykite (arsenical copper, arsenide of copper).

Copper, arsenic.

19. Enargite (sulph-arsenite of copper).

Copper, arsenic, sul-phur.

20. Harrisite(sulphide of copper)

Copper, sulphur.

21. Malachite (mountain green, green carbonate of copper, green malachite, green copper).

Copper, carbon, oxygen (water).

22. Melaconite (black oxide of copper, black copper, cupric oxide).

Copper, oxygen.

23. Native copper (sometimes with silver).

Copper (silver).

24. Pseudomalachite (phosphate of copper).

Copper, oxygen, phosphorus (water).

25. Stromeyerite (sulphuret of silver and copper, silver-copper glance).

Copper, silver, sulphur.

26. Tennantite (sulph-arsenite of copper).

Copper, arsenic, sulphur (iron).

27. Tetrahedrite (gray copper ore, sulphide of copper and antimony with various other sulphides).

Copper, antimony, sulphur (arsenic, bismuth, silver, mercury, zinc, etc)

28. Torbernite (copper-uranite, phosphate of uranium and copper).

Copper, uranium, phosphorus, oxygen (water).

29. Uranochalcite (oxide of uranium with oxide of copper and sulphate of lime)

Copper, uranium, oxygen, sulphur, calcium (water).

30. Vauquelinite (chromate of copper and lead). — Copper, lead, chromium, oxygen.
31. Whitneyite (arsenide of copper). — Copper, arsenic.

LEAD.

LIST OF THE PRINCIPAL LEAD MINERALS FOUND IN THE UNITED STATES.

NAME.	COMPOSITION.
1. Alaskaite (sulphide of bismuth, silver and lead).	Lead, bismuth, silver, copper, sulphur.
2. Altaite (telluride of lead).	Lead, tellurium.
3. Anglesite (lead-vitriol, sulphate of lead).	Lead, oxygen, sulphur.
4. Boulangerite (sulphide of lead and antimony).	Lead, antimony, sulphur.
5. Bournonite (triple sulphuret of copper, lead and antimony).	Lead, copper, antimony, sulphur.
6. Caledonite (cupreous-sulphato-carbonate of lead).	Lead, carbon, copper, oxygen, sulphur.
7. Cerussite (white lead ore, carbonate of lead).	Lead, carbon, oxygen.
8. Dechenite (vanadate of lead and zinc).	Lead, vanadium, zinc, oxygen.
9. Descloizite (vanadate of lead).	Lead, vanadium, oxygen.

10. Freieslebenite (antimonial sulphide of silver and lead).	Lead, silver, antimony, sulphur.
11. Galenite (galena, sulphide or sulphuret of lead).	Lead, sulphur.
12. Geocronite (sulph-arseno-antimonite of lead).	Lead, antimony, arsenic, sulphur.
13. Henryite (telluride of lead, with a little iron).	Lead, tellurium, iron.
14. Jamesonite (sulph-antimonite of lead).	Lead, antimony, sulphur (iron).
15. Kobellite (sulphide of lead, bismuth and antimony).	Lead, bismuth, antimony, sulphur.
16. Lanarkite (sulphato-carbonate of lead).	Lead, carbon, oxygen, sulphur.
17. Leadhillite (sulphato-tri-carbonate of lead).	Lead, carbon, oxygen, sulphur.
18. Massicot (plumbic ochre, yellow oxide of lead).	Lead, 1 part; oxygen, 1 part.
19. Mimetite (green lead ore, arsenate of lead).	Lead, arsenic, oxygen (chlorine, phosphorus).
20. Minium (red oxide of lead).	Lead, 3 parts; oxygen, 4 parts.
21. Müllerite (telluride of gold, silver and lead).	Lead, gold, silver, tellurium.

22. Nagyagite (black tellurium, foliated tellurium, telluride of gold and lead).	Lead, gold, tellurium (antimony, sulphur).
23. Native lead.	Lead.
24. Plumbogummite (phosphate of alumina and lead).	Lead, aluminum, oxygen, phosphorus.
25. Pyromorphite (phosphate and chloride of lead).	Lead, phosphorus, oxygen, chlorine.
26. Schapbachite (bismuth-silver, sulphide of bismuth, silver and lead).	Lead, bismuth, silver, sulphur.
27. Schirmerite (same as above but proportions varying).	Lead, bismuth, silver, sulphur.
28. Stolzite (tungstate of lead).	Lead, tungsten, oxygen.
29. Vauquelinite (chromate of copper and lead).	Lead, copper, chromium, oxygen.
30. Wulfenite (yellow lead ore, yellow lead-spar, molybdate of lead).	Lead, molybdenum, oxygen.

Note.—For descriptions of the above, and of other American and foreign minerals of gold, silver, copper and lead, consult Dana's System of Mineralogy, 5th Ed. with Sup., and Prof. J. Alden Smith's Report as State Geologist of Colorado, for 1880.

LIST OF USEFUL BOOKS ON SUBJECTS MORE OR LESS
CONNECTED WITH ASSAYING.

General Science.

Johnson's New Universal Cyclopædia. 4 vols. Vols. 1
and 2, 1876; vol. 3, 1877; vol. 4, 1878. New York.

General Chemistry.

Watts, H.: A Dictionary of Chemistry. 10 vols. Vols.
1–6, 1868; 1st sup., 1872; 2d sup. (vol. 7), 1875; 3d
sup. (vol. 8), Part I, 1879; Part II, 1881. London.

Chemical Technology.

Wagner, R.: A Hand-book of Chemical Technology.
Translated from 8th German edition by William
Crookes. New York, 1872.

Reference Books on Chemistry.

Roscoe, H. E., and *C. Schorlemmer :* A Treatise on Chemistry. 2 vols. Vol. 1, The Non-Metallic Elements,
1878; vol. 2, Metals, Parts I and II, 1879. New
York.

Miller, W. A.: Elements of Chemistry, Theoretical
and Practical. 3 vols. 6th edition. London,
1877–1880.

Text-books on Theoretical Chemistry.

Barker, Geo. F.: A Text-book of Elementary Chemistry, Theoretical and Inorganic. Louisville.

Roscoe H. E. : Lessons in Elementary Chemistry, Inorganic and Organic. New edition. London, 1880.

General Qualitative Analysis.

Fresenius, C. R. : Manual of Qualitative Chemical Analysis. 9th English edition. London, 1876.

Douglas, S. H., and *A. B. Prescott :* Qualitative Chemical Analysis. 3d edition. New York, 1880.

Eliot, C. W., and *F. H. Storer :* A Compendious Manual of Qualitative Chemical Analysis. New York, 1879.

General Quantitative Analysis.

Fresenius, C. R.: Manual of Quantitative Chemical Analysis. 7th English edition. London, 1876.

Classen, A.: Elementary Quantitative Analysis. Translated by E. F. Smith. Phila., 1878.

Cairns, F. A.: A Manual of Quantitative Chemical Analysis for the Use of Students. New York, 1880.

Special Quantitative Analysis.

Rammelsberg, C.: Guide to a Course of Quantitative Chemical Analysis, especially of Alloys, Minerals and Furnace Products. Translated by J. Towler. New York, 1872.

Wöhler, F.: Hand-book of Mineral Analysis. Phila., 1870.

Volumetric Analysis.

Sutton, F.: A Systematic Hand-book of Volumetric Analysis. 4th edition. London, 1882.

Hart, Edward : A Hand-book of Volumetric Analysis. New York, 1878.

Laboratory Manipulation.

Morfit (Campbell and Clarence): Chemical and Pharmaceutical Manipulations. Phila., 1857.

Williams, C. G.: A Hand-book of Chemical Manipulation. London, 1857; supplement, 1879.

Geology.

Cotta, Bernh. v. Treatise on Ore Deposits. Translated from 2d German edition by F. Prime, and revised by author. New York, 1870.

Dana, J. D.: A Text-book of Geology. 2d edition. New York, 1874.

Dana, J. D.: Manual of Geology. New York, 1881.

Le Conte, Joseph : Elements of Geology. A text-book for colleges and for the general reader. New York, 1878.

Rutley, C. L.: The Study of Rocks. 2d edition. New York, 1880.

Mineralogy.

Dana, J. D.: A System of Mineralogy. 5th edition, 1868; Appendix I, 1872; Appendix II, 1875; Appendix III, 1882. New York.

Dana, J. D.: Manual of Mineralogy and Lithology. 3d edition. New York, 1878.

•*Brush, G. J.:* Manual of Determinative Mineralogy, with an Introduction on "Blow-pipe Analysis." New York, 1878.

Foye, J. C.: Tables for the Determination, Description, and Classification of Minerals. Chicago, 1882.

Frazer, P.: Tables for the Determination of Minerals. Phila., 1874.

Hanks, H. G.: Fourth Annual Report of the State Mineralogist of California. Sacramento, 1884.

Blow-pipe Analysis.

Plattner's Manual of Qualitative and Quantitative Analysis with the Blow-pipe. Translated by H. B. Cornwall. 4th edition. Revised and corrected. New York, 1880.

Cornwall, H. B.: Manual of Blow-pipe Analysis, Qualitative and Quantitative. With a Complete System of Determinative Mineralogy. New York, 1882.

Atwood, Geo.: Practical Blow-pipe Assaying. New York, 1881.

Plympton, G. W.: The Blow-pipe. A Guide to its Use in the Determination of Salts and Minerals. New York, 1874.

Elderhorst, Wm.: Manual of Qualitative Blow-pipe Analysis. Revised by H. B. Nason. Phila., 1881.

Ross, W. A.: The Blow-pipe in Chemistry, Mineralogy, and Geology. London, 1884.

Ross, W. A.: Pyrology, or Fire Chemistry. London, 1875.

Metallurgy and Mining.

Kerl, Prof.: Practical Treatise on Metallurgy. Translated by Wm. Crookes and E. Röhrig. Vol. 1, Lead, Silver, Zinc, etc., 1868; Vol. 2, Copper and Iron, 1869; Vol. 3, Steel, Fuel, and Supplement, 1870. London.

Lock, A. G.: Gold; Its Occurrence and Extraction. London and New York, 1882.

Percy, John: Metallurgy. The Art of Extracting Metals from their Ores. Part I, Silver and Gold. London, 1880.

Percy, John: The Metallurgy of Lead. London, 1870.

Percy, John: The Metallurgy of Fuel, Wood, Peat, Coal, Charcoal, Fire-clays. Revised edition.

Callon, J.: Lectures on Mining. 3 vols. London and Paris. 1876–81.

Lamborn, R. H.: Metallurgy of Copper. 6th edition. London, 1875.

Lamborn, R. H.: Metallurgy of Silver and Lead. 6th edition. London, 1878.

Kustel, G.: Roasting of Gold and Silver Ores, and the Extraction of their Respective Metals without Quicksilver. New edition (2d). San Francisco, 1880.

Makin, G. H.: A Manual of Metallurgy. 2d edition. London, 1873.

Collins, J. H.: A First Book of Mining and Quarrying. London, 1872.

Bowie, Aug. J., Jr.: A Practical Treatise on Hydraulic Mining in California. New York, 1885.

Davies, D. C.: A Treatise on Metalliferous Minerals and Mining. 2d edition. London, 1881.

Davies, D. C.: A Treatise on Earthy and other Minerals and Mining. London, 1884.

Egleston, T.: Metallurgy of Gold, Silver, and Mercury in the United States. London and New York, 1886.

Phillips, J. A.: Mining and Metallurgy of Gold and Silver. London, 1867.

Phillips, J. A.: A Treatise on Ore Deposits. London, 1884.

Phillips, J. A.: Elements of Metallurgy. London, 1874.

Pomeroy, H. R.: Mining Manual for Prospectors, Miners, and Schools. 3d edition. St. Louis, 1881.

Kunhardt, W. B.: The Practice of Ore Dressing in Europe. New York, 1884.

Randall, P. M.: The Quartz Operator's Hand Book. Revised and enlarged. New York, 1871.

Van Wagenen, T. F.: Manual of Hydraulic Mining. For the Use of the Practical Miner. New York, 1880.

Assaying.

Aaron, C. H.: Assaying In Three Parts. Part I, Gold and Silver Ores, 1884; Parts II and III, Gold and Silver Bullion, Lead, Copper, etc., 1885. San Francisco.

Balling, C. A. M.: Die Probirkunde des Eisens und der Brennmaterialen. Prag, 1868.

Balling, C. A. M.: Die Probirkunde. Anleitung zur Vornahme docimastischer untersuchungen der Berg-und Hütten producte. Braunschweig, 1879.

Bodeman, Th., and *Bruno Kerl:* Anleitung zur Berg- und Hüttenmännischen Probirkunde. 2d edition. Clausthal, 1857.

Bodeman, Th., and *Bruno Kerl:* A Treatise on the Assaying of Lead, Copper, Silver, Gold and Mercury. Translated by W. A. Goodyear. New York, 1865.

Chapman, E. J.: Practical Instructions for the Determination by Furnace Assay of Gold and Silver in Rocks and Ores. Toronto, Can., 1881.

Kerl, Bruno: Metallurgische Probirkunst. 2d edition. Leipsig, 1882.

Kerl, Bruno: The Assayer's Manual. Translated by William T. Brannt, edited by William H. Wahl. Philadelphia and London, 1883.

Lieber, O. M.: The Assayer's Guide. Phila., 1852.

Mitchell, John: A Manual of Practical Assaying. Edited by Wm. Crookes. 6th edition. New York, 1888.

North, Oliver: The Practical Assayer. London, 1874.

Overman, F.: Practical Mineralogy, Assaying. and Mining. Phila., 1851.

Phillips, J. S.: The Explorers' and Assayers' Companion. San Francisco, 1879.

Ricketts, P. de P.: Notes on Assaying and Assaying Schemes. New York, 1879.

Silversmith, J.: A Practical Hand-Book for Miners, Metallurgists, and Assayers. New York, 1866.

Triplett, Frank: How to Assay. St. Louis, 1881*.

Metric System ; Weights and Measures.

Barnard, F. A. P.: The Metric System of Weights and Measures. 2d edition. New York, 1872.

Egleston, T.: Tables of Weights, Measures, Coins, etc. New York, 1871.

Oldberg, O.: Weights, Measures, and Specific Gravity. Chicago, 1885.

Mining Law.

Copp, H. N.: American Mining Code. 3d edition. 1880.

*Besides the above, there have been a number of publications from 1741 to about 1850 which are either obsolete in their teachings, or the information contained therein is embraced in the preceding.

Carpenter, M. B.: Mining Code. 3d edition. 1880.

Wade, W. P.: Manual of Mining Law. St. Louis, 1882.

Wilson, C. S.: Mining Laws of the United States, Colorado, New Mexico, and Arizona. 1881.

It is not pretended that the above list is complete, nor even that it comprises all the best works; it is simply a list of some that are considered standard authorities in their respective lines, save perhaps in the department of assaying, where certain ones are included that are not particularly valuable.

The plan of an assay laboratory, given on the opposite page, shows a simple and convenient arrangement, which can be adapted to almost any room.

Window

Table | Grinding Plate

Mortar Block

Working Table

LABORATORY

Assay Furnace

Window

Window

Coke or Charcoal Bin

Closet

Window

Water

Desk for Chemical Work

Door

Door

Water

Cabinet

Book Case

Window

Door

OFFICE

Window

Desk,

Scale Shelf

Window

Window

FORM FOR CERTIFICATE OF ASSAY.

Almost every assayer has his own particular blank, but so long as the certificate states plainly the results of his work, any little differences of detail are unimportant. The form given below is about as satisfactory as any.

ASSAY OFFICE, CHICAGO, ILL.,

................., 188...

I HEREBY CERTIFY that the Samples of Ores herein described, and assayed for .... gave the following results:

No.	Description.	Gold, No. of ounces per ton.	Value at $--- per oz.	Silver, No. of ounces per ton.	Value at $--- per oz.	Total Value of Gold and Silver.	Copper, per cent.	Lead, per cent.	Remarks.

.... Assayer.

ASSAYER'S OUTFIT.

With the following outfit the assayer can perform the ordinary crucible and scorification assays of gold, silver, copper and lead ores:

Hammers, sledge, medium and small, . . $	2 00
Iron mortar (8 in. diam., 1 gal.) and pestle,	1 50
Plate and Rubber, ...	10 00
Steel spatulas, one large and one small, .	75
Sieves, 20, 40 and 100 mesh, brass frames, .	3 50
Ore or pulp scales, ...	22 00
Assay balance, with weights (1 grm down to	
$\frac{1}{10}$ mgrm), ...	73 50
Set gramme weights, 100 grms down, . .	6 00
Set assay ton weights, ...	6 00
Furnace, Brown's portable, new form, .	20 00
6 muffles, size J, 12x6x4 inches, . . .	6 00
1 pair crucible tongs, ...	1 25
1 pair scorifier tongs, ...	1 00
1 pair cupel tongs, ...	1 00
Shovel, scraper and hoe, ...	1 00
Scorification mould, ...	1 00
Crucibles, 6 doz., 1 doz. covers, . . .	4 75
Scorifiers, 200 2¼ inch, 200 3 inch, . .	7 20
Cupels, 6 doz. 1¼ inch, ...	3 00
Cupel mould, 1¼ inch, brass, . . .	2 50

Piece rubber cloth, ... $ 1	00
Alcohol lamp, ...	50
Ring stand, ...	75
Wire triangle, ...	10
1 doz. 1 inch porcelain crucibles, . . . 1	44
1 doz. 1¼ inch porcelain crucibles, . . 2	16
1 quart wash-bottle, ...	75
1 pair 3 inch watch-glasses, . . .	30
Blow-pipe, ...	20
Magnifying glass, pocket size, . . . 1	00
Magnet, ...	20
Small steel hammer and anvil, . . . 1	00
Pair steel pincers, ...	25
Small cold chisel, ...	35
Horn spoon, ...	25
1 button brush	50
6 parting flasks, ... 1	20
6 annealing cups, ...	60
24 sample bottles and corks, . . .	75
12 test-tubes, assorted, ...	35
1 box gummed labels, ...	15
1 lb. bottle pure nitric acid, . . .	40
2 lbs. bi-carbonate soda, ...	20
1 lb. carbonate of potash, in bottle, . .	30
¼ lb. cyanide of potash, in bottle, . . .	25
¼ lb. borax glass, ...	25

2 lbs. flour, ... $		20
1 lb. argol, ...		15
2 lbs. nitre (nitrate potash), ...		30
2 lbs. litharge, ...		25
1 lb. charcoal, pulverized, ...		25
1 lb. silica, ...		10
½ lb. sheet lead, ...		20
1 lb. granulated lead, ...		25
2 lbs. bone-ash, ...		30
½ oz. pure silver foil, ...		75
Total, ...		$191 90

BLOW-PIPE OUTFIT.

Apparatus for Blow-piping, according to Prof. Plattner, the whole in elegant velvet-lined, polished mahogany case, with handle and lock, for travelling, complete, with case, $38. The set includes:

1 set of three porcelain dishes.

1 diamond steel mortar.

1 pair platinum pointed forceps.

1 pair heavy tip steel forceps.

1 pair steel forceps.

1 steel chisel.

1 charcoal borer, 4 cornered, with spatula.

1 charcoal borer, club shape.

1 pair fine scissors.

1 wire holder, with 3 platinum wires in the handle.

1 Plattner's blow-pipe.

lamp, with patent swivel, nickel-plated.

1 charcoal saw.

1 holder for the matrasses.

1 nickel-plated Plattner's blow-pipe.

1 heavy platinum tip for same.

1 steel hammer with wire handle.

1 set mould and stamps.

1 pair of steel nippers, Plattner's.

1 double lens.

1 knife, ivory handle.

1 dropping pipette.

1 camel's hair brush.

6 matrasses.

1 alcohol lamp, with nickel-plated air-tight top.

1 chamois skin.

6 glass tubes.

6 pieces square-cut charcoal.

Metal trays for coal, ashes, and filters.

18 flat-top, stoppered and labelled re-agent bottles, containing the following re-agents:

Test lead.

Tin.

Phosphorus salt.

Borax powder.

Borax glass.

Boracic acid, fused.

Boracic acid, cryst.

Plattner's flux.

Bismuth flux.

Carbonate soda.

Potash oxalate.

Salt.

Soda nitrate.

Charcoal.

Boneash, sieved.

Boneash, washed.

Copper oxide.

Bi-sulphate potash.

Test papers.

SECTION III.

TABLES.

MULTIPLICATION TABLE FOR GOLD AND SILVER.

SILVER.		GOLD.	
OUNCES.	VALUE.	OUNCES.	VALUE.
1........	$1 29	1......	$20 67
2........	2 58	2.......	41 34
3.......	3 87	3.......	62 01
4.......	5 16	4.......	82 68
5.......	6 45	5.......	103 35
6.......	7 74	6.......	124 02
7.......	9 03	7.......	144 69
8.......	10 32	8.......	165 36
9.......	11 61	9.......	186 03

NOTE —The above table is more relative than actual. $20.00 is commonly used as a factor for gold, and for silver the value per ounce fluctuates with the market.

TABLE OF VALUES OF GOLD AND SILVER.

WEIGHT.	Of Gold is worth	Of Silver is worth
1 grain Troy..............	$0.0430	$0.0026
1 pennyweight Troy = 24 grains Troy..............	1.0335	0.0646
1 ounce Troy = 20 pennyweights Troy=480 grains Troy	20.6718	1.2929
1 ounce Avoirdupois = 437½ grains Troy...............	18.8415	1.1784
1 pound Troy = 12 ounces Troy = 240 pennyweights Troy = 5,760 grains Troy.........	248.0620	15.5151
1 pound Avoirdupois = 16 ounces Avoirdupois = 7,000 grains Troy..............	301.4642	18.8551
1 ton Avoir. = 2,000 pounds Avoir. = { 29,166 ounces Troy 32,000 ounces Avoir. } = 14,000,000 grains Troy......	602,928.4660	37,710.3846

NOTE.—The above values are figured on the basis of $20.67·per Troy oz. for gold, and $1.29 for silver. Were the factors made $20 for gold, and the fluctuation prices of the market for silver, the values given would be varied considerably.

TABLES OF WEIGHTS.

AVOIRDUPOIS WEIGHT.

16 Drams = 1 Ounce.

16 Ounces = 1 Pound.

28 Pounds = 1 Quarter.*

4 Quarters = 1 Hundred weight.

20 Hundred weight = 1 Ton of 2240 pounds.

AVOIRDUPOIS WEIGHT.

1 dram.

1 ounce = 16 drams.

1 pound = 16 ounces = 256 drams.

1 quarter = 28 pounds = 448 ounces = 7168 drams.

1 h'dwt = 4 quarters = 112 pounds = 1792 ounces = 28672 drams.

1 ton = 20 h'dwt = 80 quarters = 2240 pounds = 35840 ounces = 573440 drams.

AVOIRDUPOIS WEIGHT.

25 Pounds = 1 Quarter.

4 Quarters = 1 Hundred weight.

20 Hundred weight = 1 Ton of 2000 pounds.

AVOIRDUPOIS WEIGHT.

1 quarter = 25 pounds = 400 ounces = 6400 drams.

* In some parts of the United States.

1 h'dwt=4 quarters = 100 pounds = 1600 ounces = 25600 drams.

1 ton=20 h dwt=80 quarters=2000 pounds = 32000 ounces=512000 drams.

TROY WEIGHT.

24 Grains=1 Pennyweight.

20 Pennyweights=1 Ounce.

12 Ounces=1 Pound.

TROY WEIGHT.

1 grain.

1 pennyweight=24 grains.

1 ounce=20 pennyweights=480 grains.

1 pound=12 ounces=240 pennyweights=5760 grains.

APOTHECARIES' WEIGHT.

20 Grains=1 Scruple.

3 Scruples=1 Dram.

8 Drams=1 Ounce.

12 Ounces=1 Pound.

APOTHECARIES' WEIGHT.

1 grain.

1 scruple=20 grains.

1 dram=3 scruples=60 grains.

1 ounce=8 drams=24 scruples=480 grains.

1 pound=12 ounces=96 drams=288 scruples=5760 grains.

1 pound, Troy, $=5760$ grains.
1 pound, Apothecaries', $=5760$ grains.
1 pound, Avoirdupois, $=7000$ Troy grains.

FRENCH OR METRIC SYSTEM OF WEIGHTS.

1 Milligramme $=.001$ of a Gramme.
1 Centigramme $=.01$ "   ".
1 Decigramme $=.1$ "   "
1 Gramme $= 1$ Gramme.
1 Decagramme $= 10$ Grammes.
1 Hectogramme $= 100$ "
1 Kilogramme $= 1000$ "
1 Myriagramme $= 10000$ "

or

10 Milligrammes (mgrs) $=1$ Centigramme (cgr).
10 Centigrammes $=1$ Decigramme (dgr).
10 Decigrammes $=1$ Gramme (grm).
10 Grammes $=1$ Decagramme (dkgr).
10 Decagrammes $=1$ Hectogramme (hgr).
10 Hectogrammes $=1$ Kilogramme (kgr).
10 Kilogrammes $=1$ Myriagramme(myrgr).

1 mgr.
1 cgr. $=10$ mgrs.
1 dgr. $=10$ cgrs. $=100$ mgrs.
1 grm. $=10$ dgrs. $=100$ cgrs. $=1,000$ mgrs.
1 dkgr. $=10$ grms. $=100$ dgrs. $=1,000$ cgrs. $=10,000$ mgrs.

1 hgr.= 10 dkgrs.= 100 grms.= 1,000 dgrs.= 10,000 cgrs.=100,000 mgrs.

1 kgr.= 10 hgrs.= 100 dkgrs.= 1,000 grms.= 10,000 dgrs.=100,000 cgrs.= 1,000,000 mgrs.

1 myrgr.=10 kgrs.=100 hgrs.=1,000 dkgrs.=10,000 grms.=100,000 dgrs.=1,000,000 cgrs.= 10,000,000 mgrs.

1 gramme=15.43235 Troy grains.

EQUIVALENTS OF SOME OF THE ENGLISH AND FRENCH
WEIGHTS.*

Troy Grains.		*Grammes.*
1	=	.064798
2	=	.129597
3	=	.194396
4	=	.259195
5	=	.323994
6	=	388793
7	=	.453592
8	=	.518391
9	=	.583190

Grammes.		*Troy Grains.*
1	=	15.43235
2	=	30.86470
3	=	46.29705
4	=	61.72940
5	=	77.16175
6	=	92.59410
7	=	108.02645
8	=	123.45880
9	=	138.89115

* T. Egleston's Tables of Weights, Measures, Coins, etc., p. 24.

ASSAY TON EQUIVALENTS IN GRAMMES, TROY GRAINS, AND TROY OUNCES.

Based on 1 gramme=15.43235 Troy grains; hence 1 assay ton or 29.166 grammes=15.43235 × 29.166=450.09992 Troy grains.

Assay Tons.	Value in Grammes.	Value in Troy Grains.	Value in Troy Ounces.
0.05	1.458	22.504	....
0.10	2.916	45.009	....
0.15	4.374	67.514	....
0.20	5.833	90.019	....
0.25	7.291	112.524	....
0.30	8.749	135.029	....
0.35	10.208	157.534	....
0.40	11.666	180.039	....
0.45	13.124	202.544	....
0.50	14.583	225.049	....
0.55	16.041	247.554	....
0.60	17.499	270.059	....
0.65	18.958	292.564	....
0.70	20.416	315.069	....
0.75	21.874	337.574	....
0.80	23.333	360.079	....
0.85	24.791	382.584	....
0.90	26.249	405.089	....
0.95	27.708	427.594	....
1.00	29.166	450.099	....

ASSAY TON EQUIVALENTS — CONTINUED.

Assay Tons.	Value in Grammes.	Value in Troy Grains.	Value in Troy Ounces.
1.05	30.624	472.604	. . . . .
1.10	32.083	495.109	1.032
1.15	33.541	517.614	1.078
1.20	34.999	540.119	1.125
1.25	36.458	562.624	1.173
1.30	37.916	585.129	1.219
1.35	39.374	607.634	1.266
1.40	40.833	630.139	1.313
1.45	42.291	652.644	1.360
1.50	43.749	675.149	1.407
1.55	45.208	697.654	1.453
1.60	46.666	720.159	1.500
1.65	48.124	742.664	1.547
1.70	49.583	765.169	1.594
1.75	51.041	787.674	1.641
1.80	52.499	810.179	1.667
1.85	53.958	832.684	1.735
1.90	55.416	855.189	1.782
1.95	56.874	877.694	1.829
2.00	58.333	900.199	1.875

ASSAY TON EQUIVALENTS — CONTINUED.

Assay Tons.	Value in Grammes.	Value in Troy Grains.	Value in Troy Ounces.
2.05	59.791	922.704	1.922
2.10	61.249	945.209	1.969
2.15	62.708	967.714	2.016
2.20	64.166	990.219	2.063
2.25	65.624	1012.724	2.110
2.30	67.083	1035.229	2.157
2.35	68.541	1057.734	2.204
2.40	69.999	1080.239	2.250
2.45	71.458	1102.744	2.297
2.50	72.916	1125.249	2.344
2.55	74.374	1147.754	2.391
2.60	75.833	1170.259	2.438
2.65	77.291	1192.764	2.485
2.70	78.749	1215.269	2.531
2.75	80.208	1237.774	2.579
2.80	81.666	1260.279	2.626
2.85	83.124	1282.784	2.672
2.90	84.583	1305.289	2.719
2.95	86.041	1327.794	2.766
3.00	87.499	1350.299	2.813

ASSAY TON EQUIVALENTS — CONTINUED.

Assay Tons.	Value in Grammes.	Value in Troy Grains.	Value in Troy Ounces.
3.05	88.958	1372.804	2.860
3.10	90.416	1395.309	2.905
3.15	91.874	1417.814	2.954
3.20	93.333	1440.319	3.001
3.25	94.791	1462.824	3.048
3.30	96.249	1485.329	3.094
3.35	97.708	1507.834	3.141
3.40	99.166	1530.339	3.188
3.45	100.624	1552.844	3.235
3.50	102.083	1575.349	3.282
3.55	103.541	1597.854	3.329
3.60	104.999	1620.359	3.376
3.65	106.458	1642.864	3.423
3.70	107.916	1665.369	3.470
3.75	109.374	1687.874	3.516
3.80	110.833	1710.379	3.563
3.85	112.291	1732.884	3.610
3.90	113.749	1755.389	3.657
3.95	115.208	1777.894	3.704
4.00	116.666	1800.399	3.751

ASSAY TON EQUIVALENTS — CONTINUED.

Assay Tons.	Value in Grammes.	Value in Troy Grains.	Value in Troy Ounces.
4 05	118.124	1822.904	3.798
4.10	119.583	1845.409	3.845
4.15	121.041	1867.914	3.891
4.20	122.499	1890.419	3.938
4.25	123.958	1912.924	3.985
4.30	125.416	1935.429	4.032
4.35	126.874	1957.934	4.079
4.40	128.333	1980.439	4.126
4.45	129.791	2002.944	4.173
4.50	131.249	2025.449	4.220
4.55	132.708	2047.954	4.267
4.60	134.166	2070.459	4.313
4.65	135.624	2092.964	4.360
4.70	137.083	2115.469	4.407
4.75	138.541	2137.974	4.454
4.80	139.999	2160.479	4.500
4.85	141.458	2182.984	4.548
4.90	142.916	2205.489	4.595
4.95	144.374	2227.994	4.642
5.00	145.833	2250.499	4.689

ASSAY TON EQUIVALENTS — CONTINUED.

Assay Tons.	Value in Grammes.	Value in Troy Grains.	Value in Troy Ounces.
5.05	147.291	2273.004	4.735
5.10	148.749	2295.509	4.782
5.15	150.208	2318.014	4.829
5.20	151.666	2340.519	4.876
5.25	153.124	23 3.024	4.923
5.30	154.583	2385.529	4.970
5.35	156.041	2408.034	5.017
5.40	157.499	2430.539	5.064
5.45	158.958	2453.044	5.111
5.50	160.416	2475.549	5.157
5 55	161.874	2498 054	5.204
5.60	163.333	2520.559	5.251
5.65	164.791	2543.064	5.298
5.70	166.249	2565.569	5.345
5.75	167.708	2588.074	5.392
5.80	169.166	2610.579	5.439
5.85	170.624	2633.084	5.486
5.90	172.083	2655.589	5.532
5.95	173.541	2678.094	5.579
6.00	174.999	2700.579	5.626

ASSAY TON EQUIVALENTS — CONTINUED.

Assay Tons.	Value in Grammes.	Value in Troy Grains.	Value in Troy Ounces.
6.05	176.458	2723.084	5.673
6.10	177.916	2745.589	5.720
6.15	179.374	2768.094	5.767
6.20	180.833	2790.599	5.814
6.25	182.291	2813.104	5.861
6.30	183.749	2835.609	5.908
6.35	185.208	2858.114	5.954
6.40	186.666	2880.619	6.001
6.45	188.124	2903.124	6.048
6.50	189.583	2925.629	6.095
6.55	191.041	2948.134	6.142
6.60	192.499	2970.639	6.189
6.65	193.958	2993.144	6.236
6.70	195.416	3015.649	6.283
6.75	196.874	3038.154	6.329
6.80	198.333	3060.659	6.376
6.85	199.791	3083.164	6.423
6.90	201.249	3105.669	6.470
6.95	202.708	3128.174	6.517
7.00	204.166	3150.679	6.564

ASSAY TON EQUIVALENTS — CONTINUED.

Assay Tons.	Value in Grammes.	Value in Troy Grains.	Value in Troy Ounces.
7.05	205.624	3173.184	6.611
7.10	207.083	3195.689	6.658
7.15	208.541	3218.194	6.705
7.20	209.999	3240.699	6.751
7.25	211.458	3263.204	6.798
7.30	212.916	3285.709	6.845
7.35	214.374	3308.214	6.892
7.40	215.833	3330.719	6.939
7.45	217.291	3353.224	6.986
7.50	218.749	3375.729	7.033
7.55	220.208	3398.234	7.080
7.60	221.666	3420.739	7.127
7.65	223.124	3443.244	7.173
7.70	224.583	3465.749	7.220
7.75	226.041	3488.254	7.267
7.80	227.499	3510.759	7.314
7.85	228.958	3533.264	7.361
7.90	230.416	3555.769	7.408
7.95	231.874	3578.274	7.455
8.00	233.333	3600.779	7.502

ASSAY TON EQUIVALENTS—CONTINUED.

Assay Tons.	Value in Grammes.	Value in Troy Grains.	Value in Troy Ounces.
8.05	234.791	3623.284	7.549
8.10	236.249	3645.789	7.595
8.15	237.708	3668.294	7.642
8.20	239.166	3690.799	7.689
8.25	240.624	3713.304	7.736
8.30	242.083	3735.809	7.783
8.35	243.541	3758.314	7.830
8.40	244.999	3780.819	7.877
8.45	246.458	3803.324	7.924
8.50	247.916	3825.829	7.970
8.55	249.374	3848.334	8.017
8.60	250.833	3870.839	8.064
8.65	252.291	3893.344	8.111
8.70	253.749	3915.849	8.158
8.75	255.208	3938.354	8.205
8.80	256.666	3960.859	8.252
8.85	258.124	3983.364	8.299
8.90	259.583	4005.869	8.346
8.95	261.041	4028.374	8.392
9.00	262.449	4050.879	8.439

ASSAY TON EQUIVALENTS — CONTINUED.

Assay Tons.	Value in Grammes.	Value in Troy Grains.	Value in Troy Ounces.
9 05	263.958	4073.384	8.486
9.10	265.416	4095.889	8.533
9.15	266.874	4118.394	8.580
9.20	268.333	4140.899	8.627
9.25	269.791	4163.404	8.674
9.30	271.249	4185.909	8.721
9.35	272.708	4208.4 4	8.768
9 40	274.166	4230.919	8.814
9.45	275.624	4253.424	8.861
9.50	277.083	4275.929	8.908
9.55	278.541	4298.434	8.955
9.60	279.999	4320.939	9.002
9.65	281.458	4343.444	9.049
9.70	282.916	4365.949	9.096
9.75	284.374	4388.454	9.143
9.80	285.833	4410.959	9.189
9.85	287.291	4433.464	9.236
9.90	288.749	4455.969	9.283
9.95	290.208	4478.474	9.330
10.00	291.666	4500.979	9.377

INDEX.

A

466; muriate, 466; native, 318, 465, 468; native, assay, 320; occurrence, 318; ores, 441; oxides, 315, 318; oxides, assay, 315, 321; oxy chloride, 466; phosphate, 468; purple, 466; pyrites, 199, 246, 263, 313, 441, 467; pyritous, 467; red oxide, 318, 441, 467; silica, 467; spatulas, 42; sulph-arsenite, 467, 468; sulphate, 466, 467; sulphides or sulphurets, 246, 312, 319, 467; sulphides, assay, 321; tests for, 424; uranite, 468; variegated, 246, 466; vitreous, 246, 312, 467; vitriol, 467; volumetric analysis, 357.

COPPERAS, re-agent, 163.

COPP'S MINING CODE, 436, 479.

CORBIN, H. H., reference to, 191.

"CORNETS," 408.

CORN STARCH, reducing power of, 147.

"CORNUCOPIAS," lead, 405.

CORNWALL'S BLOW-PIPE ANALYSIS, 475.

CORROSIVE SUBLIMATE, re-agent, 162.

CORUNDUM, mineral, 441.

"CORUSCATION," 219.

COUNTERPOISING, 411.

COVELLITE, mineral, 313, 467.

CREAM OF TARTAR, re-agent, 141, 146, 236; reducing power, 147, 175.

CRUCIBLE, process, 185, 194, 231.

CRUCIBLE TONGS, 98.

CRUCIBLES, alumina, 109; Battersea, 112; black lead (graphite, or plumbago,) charcoal-lined, 109; clay, 110; "Colorado," 112; Denver Fire Clay Co.'s, 111, 112; French clay or "Beaufay," 109, 111; gold, 109; "Gramme," 112; Hessian, iron, magnesia, platinum, 109; porcelain, 109, 120; quicklime, 109; round, 111; sand, 110; silver, 110; triangular, 111.

CRUSHERS, Bosworth, 28; Gates, 26; Taylor, 24.

CRYSTALLOGRAPHY, definition, 439.

CUPEL, color tests, 204; moulds, 132; rake, 104; shovel, 104; tongs, 102.

CUPELLATION, 197, 214.

CUPELS, directions for making, 115.

CUPREOUS, bismuth, 466; oxide, 467; sulphato-carbonate of lead, 467, 469.

CUPRIC OXIDE, mineral, 468.

CUPRITE (or cuprous oxide), mineral, 315, 318, 441, 467.

CUPS, annealing, 117.

CYANIDE OF POTASH, properties, 144; re-agent, 141, 143, 160.

CYCLOPÆDIA, JOHNSON'S UNIVERSAL, 472.

CYLINDERS, measuring, 122.

RAYMOND, R. W., reference to, 436.

RE-AGENTS, dry, for assaying, 141; for analysis, 158; bottles for, 118: in scorification process. 196; testing of, 167; wet, for assaying, 155.

REALGAR, mineral, 441, 465.

RED HEMATITES, 441.

RED OXIDE OF, copper, 315, 318, 441, 467; iron, 166, 247, 311; lead, mineral, 247, 470; lead, removal of, from litharge, 152.

REDUCING POWER OF, argol, charcoal, coke, corn starch, cream of tartar, gum arabic, hard coal, laundry starch, 147; ores, 242, 256; reducing agents, 147, 167, 175; soft coal, wheat flour, white sugar, 147.

REED, S. A., references to, 6, 31, 378.

REFERENCE BOOKS ON CHEMISTRY, 472.

REFERENCES AND LISTS, 461.

REPORT ON COLORADO MINERALS, 471.

REPORT, Fourth Annual, for California, 384.

RESIDUES, gold, weighing of, 197, 229.

RETORTING in scorifiers, 368.

RETORTS, iron, 140.

RHODIUM, element, 14; mineral, 441.

RICKETTS' NOTES ON ASSAYING, 479.

RING STAND, 129.

ROASTING, carbonate of ammonia used in, 148, 277; charcoal used in, 273; dishes, 112; of ores in crucible process, 271, 273; of ores in scorification process, 197, 199; silica used in, 273.

ROCKER, 30.

RODS, glass stirring, 123.

ROLLING MILLS, hand, 134.

ROSCOE, and Schorlemmer's Chemistry, 472; Elementary Chemistry, 473.

ROSS, W. A., books on blowpipe analysis, 476.

ROUND CRUCIBLES, 111.

RUBBER, iron, 30; sheet or cloth, 125.

RUBBING PLATE, 29.

RUBIDIUM, element, 14.

RUBY, mineral, 441.

RUBY SILVER, mineral, 182, 444, 463, 464.

RUDDLE, 166.

RUNNING crucibles in fire, 287.

RUTHENIUM, element, 14.

RUTILE, mineral, 441.

RUTLEY, reference to, 434.

RUTLEY'S STUDY OF ROCKS, 474.

S

SALT, COMMON, re-agent, 142, 148.

SALTPETRE, re-agent, 148.

SALTS, Epsom, re-agent, 163; tin, 163.

466, 467; copper and antimony, 246; copper, lead and
antimony, 466, 469; determination of in an ore, 254; ferrous or iron, re-agent, 163;
iron, 200, 245, 297; iron and
arsenic, 245; iron and copper,
181, 199, 246, 313, 319; lead,
183, 246, 316, 325, 470; lead
and antimony, 469; lead, bismuth, and antimony, 470;
manganese, 245; roasting of,
252; silver, 182, 246, 462; silver
and antimony, 182, 463, 464;
silver and copper, 183, 464,
468; silver and iron, 464; silver and lead, antimonial, 463,
470; silver, antimony, and
arsenic, 182, 463, 464; tests for,
248, 249, 250, 423; zinc, 184,
245.

SULPHOCYANIDE OF POTASSI-
UM, re-agent, 163.

SULPHUR, element, 14; determination of, in pyrites, 430; re-
agent, 142, 155, 236.

SULPHURETTED HYDROGEN,
preparation, 164.

SULPHURIC ACID, re-agent, 165.

"SURCHARGE," 410.

SUTTON, references to, 365, 428,
429.

SUTTON'S VOLUMETRIC ANAL-
YSIS, 474.

"SWEET" ORES, 275.

SYLVANITE, mineral, 461, 464,
465.

SYSTEMATIC SAMPLING of ores,
36, 38, 186, 193.

T

TABLES, apothecaries' weights,
490; assay ton weights in
grammes, grains, etc., 494;
avoirdupois to troy weights,
413; avoirdupois weights, 489;
equivalents of English and
French weights, 493; French
or Metric weights, 491; multiplication for gold and silver, 487; reducing powers of
reducing agents, 147; sulphurets, 252, 263; troy, 415, 490;
troy to avoirdupois weights,
415; values of gold and silver,
417, 488; weights, 489.

TALC, mineral, 442.

TALCOSE-SCHIST, 442.

TANTALUM, element, 14.

TARTAR, cream of, re-agent, 141,
146; reducing power, 147, 175.

TARTARIC ACID, re agent, 165.

TAYLOR CRUSHER, 24.

TECHNOLOGY, CHEMICAL, WAG-
NER'S, 472.

TECHNOLOGY QUARTERLY, reference to, 433.

TELLURIC SILVER, mineral, 463.

TELLURIDES, 461; by scorification process, 202; gold, 461;
gold and lead, 461, 471; gold
and silver, 461, 463, 464; gold,
silver and lead, 461, 470; lead,

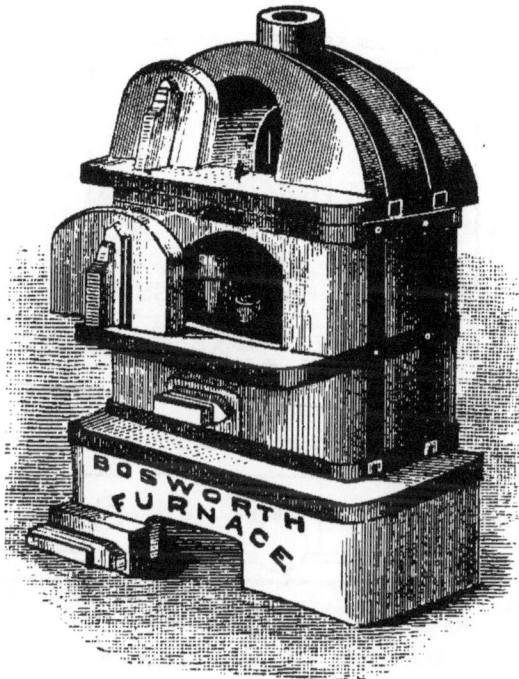

HENRY TROEMNER,

710 MARKET STREET, PHILADELPHIA.

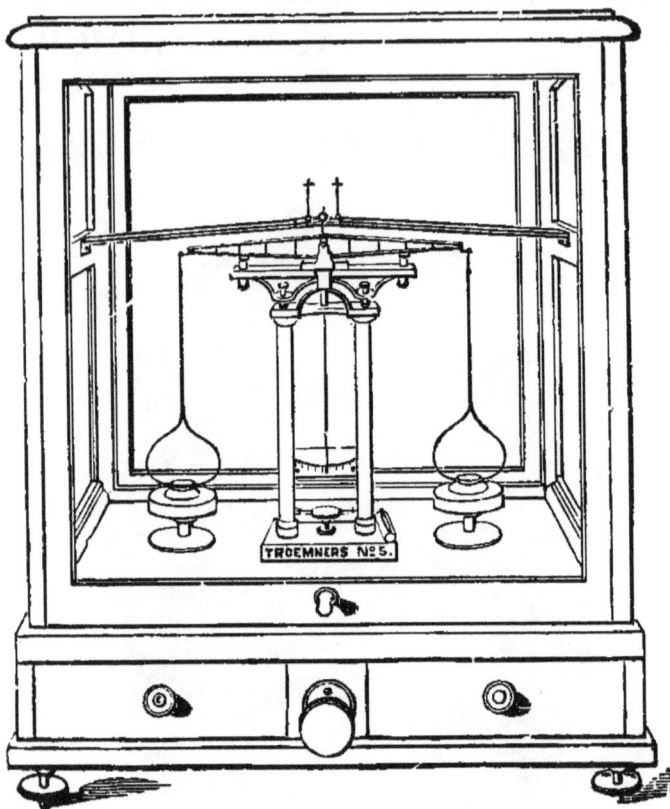

MAKER OF ACCURATE

ANALYTICAL AND ASSAY

BALANCES and WEIGHTS

In use at all the United States Mints and Assay
Offices.   A Full Description Sent on Application.

(SEE "BALL BALANCE," PAGE 49.)

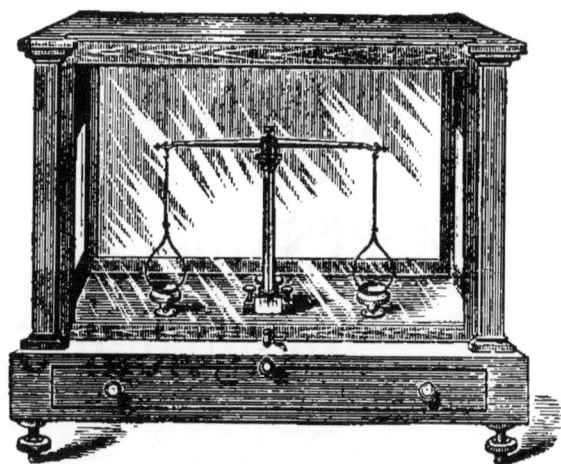